Highway Meteorology

Other titles from E & FN Spon

City Centre Planning and Public Transport
B.J. Simpson

Computer Graphics for Surveying
P.H. Milne

The Contractors Guide to the Law of Highways
R.A. O'Hara

Environmental Planning for Site Development
A.R. Beer

Fundamentals of Weather and Climate
J.F.R. McIlveen

Geomembranes
Identification and performance testing
Edited by A.L. Rollin and J.M. Rigo

A Handbook of Segmental Paving
A.A. Lilley

Laser Surveying
W.F. Price and J. Uren

The Maintenance of Brick and Stone Masonry Structures
Edited by A.M. Sowden

Mechanical Tests for Bituminous Mixes
Characterisation, design and quality control
Edited by H.W. Fritz and E. Eustacchio

Parking
A Handbook of environmental design
J. McCluskey

Remote Sensing of Ice and Snow
D.K. Hall and J. Martinec

Highway Meteorology

Edited by

A.H. Perry
and
L.J. Symons
Department of Geography,
University College, Swansea, Wales

E & FN SPON
An Imprint of Chapman & Hall
London · New York · Tokyo · Melbourne · Madras

Published by E & FN Spon, an imprint of Chapman & Hall, 2–6 Boundary Row, London SE1 8HN

Chapman & Hall, 2–6 Boundary Row, London SE1 8HN, UK

Van Nostrand Reinhold Inc, 115 5th Avenue, New York NY10003, USA

Chapman & Hall Japan, Thomson Publishing Japan, Hirakawacho Nemoto Building, 7F, 1-7-11 Hirakawa-cho, Chiyoda-ku, Tokyo 102, Japan

Chapman & Hall Australia, Thomas Nelson Australia, 102 Dodds Street, South Melbourne, Victoria 3205, Australia

Chapman & Hall India, R. Seshadri, 32 Second Main Road, CIT East, Madras 600 035, India

---

First edition 1991

Typeset in 10/12pt Palatino by Best-set Typesetter Ltd
Printed in Great Britain by St Edmundsbury Press Ltd, Bury St Edmunds, Suffolk

ISBN 0 419 15670 4   0 442 31380 2 (USA)

A catalogue record for this book is available from the British Library

Library of Congress Cataloging-in-Publication Data

Highway meteorology/edited by A.H. Perry and L.J. Symons.
p. cm.
Includes bibliographical references and index.
ISBN 0–442–31380–2
1. Climatology. 2. Transportation, Automotive—Climatic factors. 3. Roads—Safety measures. 4. Roads—Maintenance and repair. I. Perry, A.H. (Allen Howard) II. Symons, Leslie.
QC981.H5919 1991
363.12′51—dc20   91-9060
CIP

Printed on permanent acid-free text paper, manufactured in accordance with the proposed ANSI/NISO Z 39.48–199X and ANSI Z 39.48–1984

Biographies of Contributors

EDITORS

Allen Perry is senior lecturer in the Department of Geography, University College, Swansea. He is author of 'Environmental hazards in the British Isles' (Allen & Unwin 1981) and many papers on the climatology of ice and snow and its impact on transport systems. He was a member of the Climatic Change Impact Review Group convened by the Department of the Environment to consider the likely impact of global warming on the UK and its economy.

Leslie Symons is Professor of Geography at University College Swansea with a particular interest in transport systems and their operation. He has contributed papers on weather and transport to successive meetings of the Standing European Road Weather Conference (SERWEC) and has undertaken contract research on highway winter maintenance problems.

CONTRIBUTORS

Anthony J. Brazel is director of the Laboratory of Climatology at Arizona State University, Tempe, Arizona. He is an authority on the problems of blowing sand and dust in the south-western US.

Stanley L. Ring is Professor of Civil Engineering at Iowa State University, Ames, Iowa and is the author of many papers on the siting and efficiency of snow fences in the US.

James A. Nanninga is the Director of Public Works for the city of Aurora, Illinois, situated in the US 'snow-belt', where the task of clearing roads of large snowfalls is a regular winter occurrence.

Leslie Musk died under tragic circumstances while this book was in production. He was a geography lecturer at the University of Manchester specialising in the study of applied climatology and he had completed a number of fog consultancy studies for UK local authorities.

Erkki Nysten is a meteorologist working at the Finnish Meteorological Institute in Helsinki. He is a member of the Standing European Road Weather Executive Committee and chairman of the management committee of COST 309.

Jean Palutikof is the Assistant Director (Research) of the Climatic Research Unit at the University of East Anglia. Her research interests are in the area of climatic change impacts and the application of climatic data to economic and planning issues.

John Thornes is a lecturer in the School of Geography University of Birmingham. He was co-founder of Thermal Mapping International and at present is co-principal investigator of the Strategic Highways Research Programme contract investigating winter storm monitoring and communications.

Contents

Metric/Imperial Conversion Chart

Mass

1 kg = 1000 g = 2.205 lb
1 g = 1000 mg = 0.035 oz
1 tonne = 0.984 ton

1 lb = 0.454 kg
1 oz = 28.350 g
1 ton = 1.016 tonne

Length

1 km = 1000 m = 0.621 miles
1 m = 3.281 ft = 39.37 inches
1 cm = 10 mm = 0.394 inches

1 mile = 1.609 km
1 ft = 0.305 m
1 inch = 25.4 mm

Speed

1 m/s = 3.281 ft/s = 1.943 knots*
1 km/h = 0.278 m/s

1 ft/s = 0.305 m/s
1 mile/h = 1.609 km/h

Pressure

1 mb = 10^2 psi = 0.0145 lbf/in^2

*1 UK knot = 1.8532 km/h
= 1.00064 knots (international)

Acknowledgements

The authors wish to acknowledge the support of the Director of Highways, Welsh Office, for research at the University College of Swansea 1982–90 into combating snow and ice and other meteorological problems. The Meteorological Office are also thanked for their help during this project.

They also wish to thank the many individuals, too numerous to mention individually, who have assisted their researches throughout the UK, Europe and North America, during the past 15 years of involvement with problems of climate affecting transport.

Preface

It is surprising, considering the dependence of modern economies on road transport, that to the editors' knowledge this is the first book that has attempted to look at the impact of weather on highways. However, the effects of adverse weather on communications can disrupt and dislocate travel patterns and affect a wide range of industrial and commercial activity. While it might appear self-evident that the safety, journey time and indeed the feasibility of undertaking any trip can be influenced by the weather, too often the driver cocooned behind the steering wheel feels safe and immune from the elements outside. The complex characteristics of the earth's surface interact with the very lowest layers of the atmosphere to create a great variety of weather and climate conditions in both space and time. Thus sharp differences in weather can develop within short distances over many road networks, which are difficult to forecast and difficult for the road user to foresee. The weather experienced on a journey is partly due to the vehicle's motion as well as to the changing weather situation. Thus if a rainbelt is moving at 40 mile/h and a driver travelling in the same direction averages 50 mile/h he will experience inclement conditions for two or three times the duration at an individual location.

At the current rate of growth there are likely to be 400 million vehicles worldwide by the end of the century; of these 165 million will be in Europe, of which over 25 million will be in the UK. Increased pressure on road space from higher traffic densities means that safer roads and better-informed road users will become increasingly important if accident rates are not to rise dramatically. In the past twenty years the volume of traffic has increased by 92% in the UK while the length of public road has increased by only 7.5%. Spending on advanced technology aids such as ice and fog detection systems seems likely to rise. Not only are road users affected by the weather, but so numerous have they become that they are increasingly having an environmental impact and perhaps even helping to change the climate. Some exhaust products such as nitrous oxide are implicated in the increased greenhouse effect which many experts predict will raise global temperatures by up to 5 °C over the next 50 years. Pollution from vehicles amounts to about 20% of the UK national emissions of carbon dioxide (the principal greenhouse gas), and about 40% of acid rain and all airborne lead. Already the motor vehicle is a prominent contributor to air pollution and degradation of air quality, especially in cities and built-up areas. There is unlikely to be a significant cut in the total amount of greenhouse gases from motor car exhausts in the next twenty years, with improvements in engine design outweighed by the increase in the number of cars. Strict emission standards of the type already employed in Japan and the USA are spreading to Europe. The action of sunlight on the complex brew of exhaust gases is producing photochemical smogs in many sub-tropical and tropical cities. In Athens, for example, the dense cloud of pollution, or *nefos* as the Athenians call it, has resulted in restrictions being imposed on private vehicles, which are allowed into the city's inner circle only on alternate weekdays.

Toxic substances in the air ingested into the vehicle can modify the microclimate and cause physical stress to the driver. Pollutants can affect concentration, cause drowsiness, increase reaction time and so, perhaps, be a contributory cause of road accidents. There is plenty of evidence that not only are motorists extremely loath to give up using their cars, but they even resent altering their driving habits. During winter 1988/89 pollution levels increased alarmingly over central Europe under a persistent anticyclone. In Geneva, illuminated signs at busy crossroads exhorted drivers to *coupez le moteur*. While 50% of drivers heeded this request to begin with, this figure soon fell to 10%; many drivers preferred to run the engine and keep the heater running in the cold smoggy weather. There are few signs that these considerations have crossed the minds of those designing roads to combat congestion. New roads tend to create new generations of road user, since journeys are perceived as easier and quicker, and hence personal freedom of movement becomes more attractive. Global pollution problems, and the significant contribution produced by vehicles, must reopen the argument over what form of transport any country should have. A policy of satisfying all demands for roads is hardly consistent with one that exhorts power companies to cut their carbon dioxide emissions.

Because the links between weather and the road user are many and various the editors have found some difficulty in structuring the contents of this book. Inevitably perhaps the treatment has focused on highway weather hazards and the techniques that can be employed to combat them. In middle latitudes it is winter that is perceived as being likely to produce conditions that can seriously affect road surface, visibility and vehicle handling. Concern about such conditions in the UK led in 1989 to the publication by British Road Services of a winter hazards map to highlight the problems.

In Chapter 1, climate is considered as a factor in new road design. In the following six chapters detailed consideration is given to some of the weather conditions that are most likely to produce hazardous conditions for the road user. In Chapters 2–5 hazards that reduce friction between the tyre and the road surface are examined. A characteristic of almost all road vehicles is the small area of contact with the road surface. Any meteorological phenomenon that reduces the skid-resistance of the road can have a detrimental effect on road safety. The point at which cohesion between the tyre and the road surface is lost will depend on such factors as the condition of the tyres, the speed of the vehicle, the skill of the driver and the material comprising the road surface. Most important of all is likely to be the state of the road surface. The passage of traffic will deposit a film of oil and rubber on the highway which can build up during dry conditions and then form an extremely slippery surface when rain falls. Transition from reliable adhesion to virtually no adhesion typically occurs almost instantaneously either as water on the surface freezes or as other surface conditions, such as excessive wetness, promote aquaplaning.

For other forms of transport, like air and rail, the impact of ice and snow is most keenly felt at a limited number of points on the transport network, at track points (switches) and at airports for example, but on the roads the problems are more widespread and thus resources have to be spread more thinly. 'Snow and ice control' is the phrase used in North America to cover what in Europe is often called 'winter maintenance'. It seems likely that the average annual bill for such facilities in the Northern Hemisphere is at least £2000 million (Thornes 1988). Whether this is a cost-effective response to winter maintenance requirements remains in doubt, but any savings of material and labour costs which can be achieved by new advances like road sensors, must be of interest to roadmasters.

While technological changes continue to alter the face of the winter maintenance 'industry', the broad signposts to the future are now in place and are examined in detail in these chapters. Europe has taken a world lead in developing and deploying new winter-maintenance technology, probably because the prevailing ethos tends to be one of taking preventive action. In Germany, for example, it has been suggested that a rate of 1.0 accidents on a given stretch of road can be reduced to 0.83 with a well maintained winter highway. In contrast in North America where very little pre-salting takes place, more attention has been paid to improving curative actions such as snow blowing, and this is considered in Chapters 4 and 5. Only recently with the Strategic Highways Research Program (SHRP) is attention turning to ice-pavement bond prevention rather than bond destruction. This five-year programme, which began in 1987 and is costing $30 million a year, is the largest of a number of national projects concerned with highway meteorology. In France, SEMER is trying to reduce by 1% deaths and injuries caused by fog and ice on the road using satellite dissemination of weather charts, radar and satellite pictures and ice predictions which can be received by the roadmaster on the same screen using the METEOTEL system that is currently under development. In Sweden an integrated weather system providing very short-range, detailed forecasts, that will be efficiently disseminated and tailored to road users' needs, is being developed under the PROMIS 90 plan.

Chapters 6 and 7 are concerned with hazards that reduce the visibility of the highway and other vehicles to the driver and by impairing his vision influence his decision-making processes. While, as the OECD (1976) have pointed out, fog is a rare phenomenon, even in countries or regions that seem to be struck very often by fog, it is the aspect of weather that drivers most fear, probably because it is difficult to take ameliorative action. While ice and snow can cause as much disorganization these hazards can be dealt with, but there is no known means of preventing the occurrence of fog, or of dispersing it at an acceptable cost when it does occur. Though the total number of accidents which are fog-related may be small, they are generally multiple collisions, and frequently receive mass media attention, often with emotive headlines such as 'motorway madness'.

Reductions in visibility as a result of non-aqueous particles like sand and dust can be just as dangerous, as the case study from Arizona in Chapter 7 shows. There are large areas of the world where the climate is either arid or seasonally arid, and where such problems are widespread. If, as is often the case, road surfaces are unmetalled, the problem is likely to be exacerbated.

Weather and road accidents are considered in Chapter 8, while in Chapter 9, international programmes of collaboration in tackling the impact of weather on highways are reviewed.

Good progress has been made in dealing with weather hazards on highways, particularly during the 1980s, but any complacency that may build up can be shattered by an unusually bad spell of weather. In Great Britain, for example, after several mild winters, late 1990 and early 1991 saw chaos for several days on British roads, railways and airports as snow struck. Although blizzards were brief by comparison with American or Scandinavian experience, the transport services over much of England were unable to cope. The government examined arrangements made by other countries for minimizing the effects of bad weather on freight and passenger transport. One measure reported to be under consideration was the temporary closure of motorways to prevent vehicles from becoming snowed up and requiring major rescue services. The situation on the motorways in the Midlands resembled the forced closure of the M4 in Wales as long ago as 1982 when hundreds of cars and heavy goods vehicles had been snowed up for several days, many of the lorries because the diesel fuel had frozen in the tanks of the vehicles. This is one lesson that has been learned, with the specification of diesel fuel for use in Britain now protecting it down to a much lower temperature than it was in 1982 when it 'waxed' at −9°C. (Williams, Symons and Perry 1984).

REFERENCES

British Road Services (1989). *The BRS Winter Hazards Map*, BRS Western Ltd.

OECD (1976). *Adverse Weather, Reduced Visibility and Road Safety*. Paris: OECD.

Thornes, J. (1988). Towards a cost/benefit analysis of the UK Natonal Ice Prediction System. *Proceedings of the Fourth International Conference on Weather and Road Safety*, Academia dei Geogrofili, Florence, Italy, pp. 559–579.

Williams, P.J., Symons, L.J. and Perry, A.H. (1984). The waxing of diesel fuels. *Journal of Institute of Highways and Transportation*, **31**, 34–41.

A.H. Perry and L.J. Symons
Swansea

June 1991

Chapter One

Climate as a factor in the planning and design of new roads and motorways

Leslie F. Musk

The study of local weather hazards and their effects on road traffic is of ever-increasing concern in the design and planning of new roads and motorways. Accurate knowledge of the spectrum of climatic conditions which are likely to be experienced on new roads is necessary to ensure the safety of eventual road users and to ensure that the road is economically viable.

The weather influences the safety of road users in two ways. It affects the volume of traffic on the road (particularly private cars), and it may reduce the number of users exposed to the possible risk of weather-induced accidents arising from slippery roads, poor visibility, strong winds, snow and ice. When comparing a number of routes for a particular new road scheme, those routes exposed to the worst of the local climatic conditions may be expected to take significantly less traffic than the others, and furthermore the accident rate on these routes may be unacceptably higher than on the others. The more exposed routes will also have additional costs arising from: the increased costs of providing permanent protection to the road (from the spalling action of frost for example); the increased costs of protecting road traffic (using wind-breaks, snow fences, road gritting etc.); the increased construction costs arising from such factors as loss of working time and deterioration of plant; and the increased costs of resurfacing and repainting the finished road.

Increased attention is now being given to the problems of forecasting and ameliorating the problems of weather hazards on existing roads and to the minimizing of such problems on new roads at the planning stage. For existing roads and motorways the Meteorological Office provides a wide range of dedicated forecast services for road transport, including the Open Road weather service for highway authorities, concerned with winter road maintenance, launched in late 1986 (Hunt 1987). A range of new technology is now available to assist the local highway engineer, much of which is described in detail elsewhere in this book. Organizations such as Weather Watchers have developed to produce and coordinate updated information on road weather conditions in Scotland and elsewhere (Chaplain 1987), while the Standing European Road Weather Commission (SERWEC) has been established to foster international cooperation and the exchange of information on all aspects of highway meteorology.

When planning new roads and motorways, the choice of alternative routes often lies within a relatively narrow band of terrain, and climatological hazards are just one of the many factors which have to be considered when appraising a new road scheme. However, where the alternatives traverse areas of widely different climatological characteristics, for example the alternative routes for the M62 trans-Pennine motorway, and the A55 North Wales Coast Road, the weather factor can be of major significance when choosing between routes.

1.1 CLIMATOLOGICAL SURVEYS

The importance of climatological variations on the local scale was first recognized as an important factor in the planning of new motorways in the UK in the early 1960s, in connection with the planning of the M62, the trans-Pennine motorway from Liverpool to Hull. Existing routes across the Pennines, such as the Snake Pass and the Woodhead Pass, were very congested and suffered from ice, snow and fog during the winter months; on occasions this caused them to be closed completely. Several routes were suggested for the motorway, none of which exceeded 427 m (1400 ft) above sea level at any point (Lovell 1966). Although this figure is modest when compared with the altitudes dealt with by engineers designing routes through such mountainous areas as the Alps and the Himalayas, this was the first occasion on which British engineers had to address a planning problem of this nature. It was hoped that adverse weather conditions could be minimized as far as possible so that about 100 000 vehicles per day could flow freely along the route on every day of the year.

Of the routes considered, the choice ultimately lay between a high-level route (maximum height 411.5 m) and a low-level route (maximum height 365.5 m). Existing meteorological records for the area were too sparse to make useful comparisons between the two routes and, following discussions with the Meteorological Office, a network of ten weather stations was established in January 1962 in the immediate vicinity of each route. The stations, located equidistantly along the routes in question, were used to observe temperature, visibility, rainfall, snowfall, wind speed and wind direction. Although the survey was quite short (being terminated in May of the same year), a comparison of the results with records from existing meteorological stations nearby indicated that they were representative, and gave a good indication of the severity of the weather conditions which could be expected in the region. The period of the survey included part of the worst winter since the Second World War.

The high-level route experienced visibility less than 400 m for 30–40% longer than the low-level route; it experienced 20% more snow and 10% more ground frost during the survey. The climatic factor was of a scale never before experienced in planning a motorway. Weather was a deciding factor in choosing between alternative routes, and some of the engineering designs and structures were modified because of the severity of the weather. The route opened in December 1970, and there have been very few serious accidents on the high-level section of the M62 which can be attributed to the weather: a considerable improvement on the old routes.

The project received widespread publicity and acclaim. A similar investigation was subsequently conducted for the M6 motorway over Shap Fell in Cumbria, and a meteorological survey was later conducted in the Longdendale valley for a proposed motorway linking Manchester and Sheffield. Such surveys are not common, however. More recent surveys for new roads have tended to concentrate on particular meteorological parameters such as visibility, wind or snow. Climatologists are today called upon by planners and engineers to advise them on:

(a) the choice between alternative routes purely on the grounds of weather hazards;
(b) the identification of particular locations on proposed new routes where weather hazards may be a problem;
(c) the quantification of the weather hazard in terms of its severity and its likely frequency; and
(d) how the hazard might be ameliorated.

Such advice and information is subsequently used in the technical appraisal of alternative routes, public consultations and public inquiries.

The author provided such advice on fog and other climatological hazards (particularly wind) for the Department of Transport and the Welsh Office for eight new road schemes between 1975 and 1989;

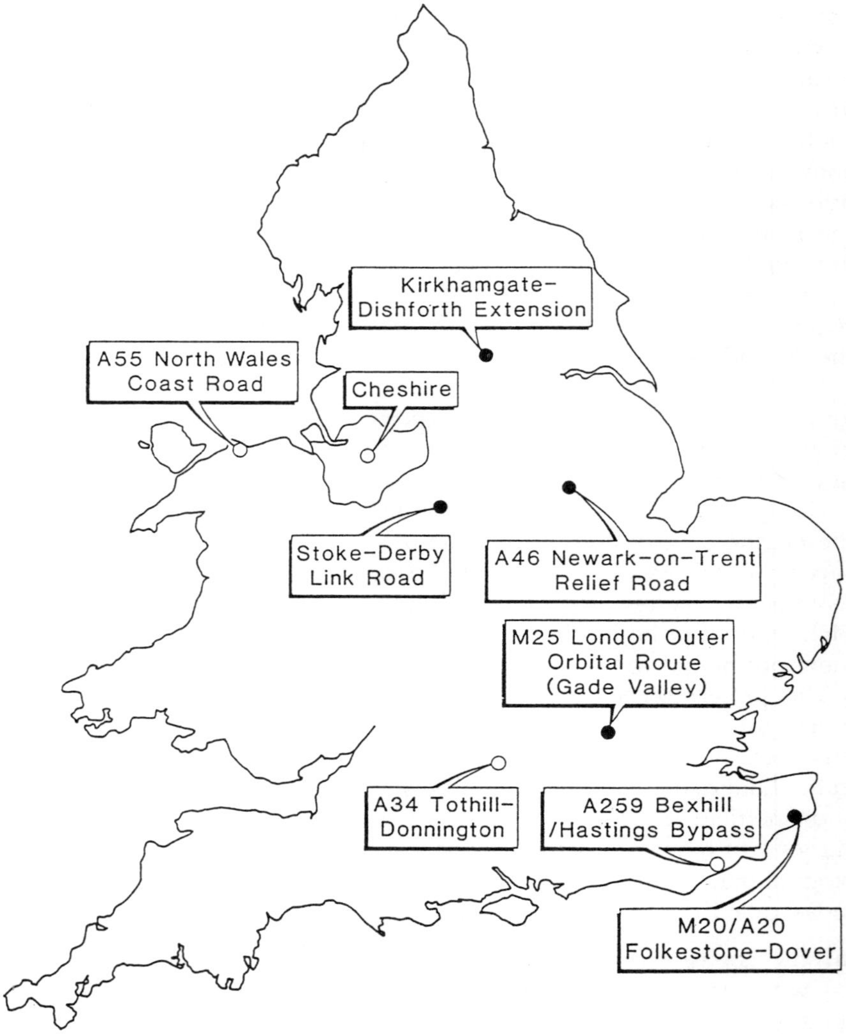

Figure 1.1 New road and motorway schemes for which climatological advice has been supplied by the author. Solid circles indicate studies for which *in-situ* field data were collected; open circles indicate desk studies.

these schemes are shown in Figure 1.1 (together with the location of a fog survey of the major roads and motorways of Cheshire). Four of these schemes involved desk studies (those shown by open circles in Figure 1.1) which integrated data from standard meteorological stations, observations from stations run by local 'amateur enthusiasts', comments from the local police, the Automobile Association, the RAC, comments from local inhabitants, analyses of weather-related road accidents in the locality, and evaluation of the variation of the Fog Potential Index (see Chapter 6) along selected routes (Musk 1982).

Five of the schemes (shown by solid circles in Figure 1.1) involved field surveys and the collection of *in-situ* data. These are listed below, but because of the degree of confidentiality involved in such work, can only be described in outline.

(a) M64 Stoke–Derby Link Road (survey completed 1976)
Data from 24 sites where visibility was observed manually (together with calculated Fog Potential Indices), were used to evaluate the local variability of fog along alternative lines for this proposed road connecting the M6 and the M1 through the Potteries.

(b) A46 Newark-on-Trent Relief Road (survey completed 1979)
Visibility boards were erected at nine sites along alternative routes for this local relief road; manual observations of these provided useful information on the effects of such local features as the River Trent, the town of Newark, Staythorpe power station, gravel workings and steam from the town's sugar beet factory, on the variability of the climatology of fog along the routes.

(c) M1/A1 Kirkhamgate–Dishforth Extension (survey completed 1980)
A fog survey, using data from twelve sites where visibility was recorded manually, four sites where it was recorded continuously by transmissometers, and calculated Fog Potential Indices, was conducted to investigate local variations in the fog climatology along a corridor where alternative routes were proposed for an extension of the M1 to the A1 on the eastern side of Leeds.

(d) M25 London Outer Orbital Motorway – Gade Valley
(survey completed 1982)
Visibility data from three transmissometers, together with an evaluation of Fog Potential Indices in the locality, were used to investigate the difference in the incidence of fog at two alternative crossings of the Gade valley in Hertfordshire. The fogginess of the locality was one factor which prompted the construction of large pillars to carry the M25 well above the valley floor, and hopefully out of serious danger from the fog which forms in the valley below.

(e) M20/A20 Folkestone–Dover (survey completed 1988)
Wind data were collected from four anemographs (and 15 tatter flag sites, but these were not used because of vandalism). Visibility data were obtained from two transmissometers and 24 visibility traverses over 12 sites. These were used to assess the severity of local weather hazards on this road which will link Dover with the Channel Tunnel terminal at Cheriton. Particular problems which were investigated were the severity of the winds over the cliffs, and the extent to which sea fog and local hill fog penetrated inland from the coast.

In the early 1970s a visibility survey was conducted in the Tapster Valley in connection with the planning of the Warwick section of the M40 Oxford to Birmingham motorway (Hogg 1973). Fifteen sites were chosen where visibility was observed manually; the survey was subsequently extended to monitor conditions at 45 locations along all the alternative routes for the motorway in the Warwick to Umberslade corridor.

In 1985, the Meteorological Office was commissioned to examine the susceptibility to fog of the M25 motorway around London, following a serious accident in dense fog on the Surrey section. Its aim was to identify stretches which are particularly prone to fog with a view to the installation of lighting, warning systems or fog detectors. The survey was based on an analysis of visibility data from standard meteorological stations, local information, traverses of the motorway, analysis of satellite data for days when fog was known to have occurred, and the use of the author's Fog Potential Index (Meteorological Office 1985). The Department of Transport has tested five different fog detecting systems and has installed selected instruments at the most fog-prone locations on the M25 London orbital motorway (see Chapter 6).

Apart from a survey of the local variation of snow and frost on alternative routes proposed for the Okehampton bypass in Devon, there have been few other climatological surveys conducted in connection with the planning of new roads. Frequently there is insufficient time available to set up and obtain reliable results from special meteorological surveys such as those quoted. In 1991 the Meteorological Office set up a Land Transport Consultancy Group to identify fog black spots and other weather hazards on projected roads.

1.2 WEATHER HAZARDS AFFECTING ROAD TRAFFIC

The most important meteorological parameters affecting road users and their safety are rainfall, snow, ice, fog and wind; their effects are exacerbated when they occur in combination and in darkness. Apart from wind, which will be discussed later in this chapter, the other

Table 1.1 *Some preventive and remedial measures which can be used to ameliorate the effects of weather hazards on roads*

Hazard	*Effects*	*Preventive measures*	*Remedial measures*
Rainfall	Poor visibility due to the rain and spray from road and vehicles	Legislation requiring effective sprayguards on vehicles espec. lorries; need for method for keeping motorcycle windshields and helmet visors free from water droplets	Use of pervious macadam as a surfacing material to prevent splash and spray
	Skidding	Good road drainage and surfacing to reduce risk	–
	Dazzle from wet pavements	–	–
	Flooding	Good drainage design	–
Snowfall	Road blockage	1. Design earthwork side-slopes to minimize drifting 2. Snow fences	Removal by snow ploughs and snow blowers Closure of road before crisis develops
	Skidding	–	1. Road gritting and salting 2. Use of studded tyres and snow chains

Table 1.1 Cont.

Hazard	*Effects*	*Preventive measures*	*Remedial measures*
Ice and frost	Skidding	Design to minimize areas in shadow; avoid known frost hollows	1. Road heating 2. Road salting and gritting
Fog	Poor visibility	All very expensive and not in widespread use: 1. Evaporation of droplets by heating 2. Promotion of condensation by seeding 3. Promotion of droplet coalescence by various methods 4. Raising height of road (on pillars) above fog-prone valleys	1. Low-level lighting 2. External vehicle lights 3. Internal warning devices 4. Police convoy systems 5. Fog warning signs 6. Speed restrictions 7. Radio and television broadcasts
Wind	Swerving and overturning	1. Design to avoid sudden transitions from sheltered to exposed areas 2. Wind-breaks	1. Speed restrictions 2. Warning signs 3. Radio and television broadcasts 4. Road closure to high-sided vehicles

hazards are all discussed in later chapters. Many of the effects of the hazards are well known. There are a variety of possible measures which can be taken to ameliorate their effects (see Table 1.1): *preventive* measures prevent or reduce the impact of the hazard on the road user; *remedial* measures help alleviate the problems once they have occurred.

1.2.1 Rainfall

In numerical terms the greatest weather hazards are associated with rain and wet roads; they occur much more frequently than fog, ice or snow. Some 20% of *all* road personal injury accidents in Great Britain are caused by skidding, mostly on wet roads, and nearly half of them when rain is actually falling. A number of studies of accident statistics in this country (Codling 1974; Smith 1982), in Australia (Robinson 1965) and in the USA (Sherretz and Farhar 1978) have shown how accident rates increase by some 30% on rainy days compared with non-rainy days.

The principal problems of rainfall for road traffic are poor visibility, loss of skidding resistance (aquaplaning) and, to a lesser extent, reflections from a wet road surface at night. Where the road is exposed to strong winds, the problems of driving rain make the situation worse. Spray, which is often dirty, thrown up by passing vehicles will cause visibility conditions to deteriorate, particularly in darkness. A very important consideration for drivers of scooters and motorcycles in rain is that they have no facility for wiping water droplets off their windshields or the visors of their helmets; with reflections from surface water on the road frequently obscuring road markings, this can be a potentially dangerous situation, especially for the inexperienced.

With heavy rainfall, flooding may pose localized problems, but the use of weather radar now allows forecasters to see where it is raining and by how much, so that prediction of the hazard can be based on good real-time information (see Chapter 2). With a good road-drainage system the problem can be greatly alleviated.

1.2.2 Hail

Hail can cause unexpected and serious problems for drivers in certain situations. Pike (1986) has described a major accident on the M4 motorway in Berkshire in November 1985 caused by hail, involving six vehicles on one carriageway and thirteen on the other. A sudden violent hailstorm produced hail 5–10 mm in diameter which covered the unprotected hard shoulder to a depth of 2–3 cm. Drivers lost control of their vehicles, in a situation similar to 'driving on ball bearings', and Pike suggests that the Highway Code should be amended to make it permissible for drivers to stop on the hard shoulder (if they can) during such severe hailstorms.

1.2.3 Snowfall

Compacted snow and snow-drifts can cause problems for the road user, and even cause road blockage. A single snowstorm, lasting for only a few hours, can cause more disruption than its magnitude might suggest. Perry and Symons (1980) have described in detail the impact of the Scottish blizzard of 28–30 January 1978 on the transport of the country.

Road casualties tend to increase by some 25% on snowy days, compared with non-snow days, but the casualty rate per unit of traffic tends to be nearly doubled (OECD 1976).

Forecasting snowfall is difficult, especially in terms of pinpointing where the snow is going to be heaviest, how much will fall and whether it will lie; a change of 1 °C can make all the difference between a wet, slushy day and one where road travel is chaotic. For traffic disruption to occur, the snow need not occur in deep drifts; a covering of 20 mm or less in thickness may be sufficient to hinder or even halt traffic (Perry and Symons, 1980).

When temperatures are low, individual snowflakes are small and have a less branching structure than in warmer temperatures; they are thus more prone to the action of wind, producing dangerous conditions, especially at night. Snowdrifts are caused by obstacles in the windstream, decreasing the wind velocity and causing snow deposition in their lee. In the 1960s field experiments and wind tunnel tests were conducted to design road cross-sections which would discourage snow-drifts from accumulating on the M62 (Lovell 1966). It was found that the optimum cross-section for motorway embankments had a side-slope of 1 in 5; this minimizes the gustiness of the wind over the embankment. As a consequence, the approach cutting of the M6 to Scammonden Dam, 11 km west of Huddersfield, was modified to alter the behaviour of snow-drifts and reduce the risk of snow avalanches from the high side-walls of the cutting.

The use of snowfences ensures the deposition of the maximum amount of snow in advance of a road (Chapter 5). Price (1971) undertook a series of experiments in Inverness which showed the need to leave a 20 cm gap between the bottom of a snow fence and the ground so that a powerful rising eddy could be formed on the leeward side of the fence.

The hazards produced by *ice*, *frost* and *fog*, and the methods by which these might be ameliorated, are summarized in Table 1.1, and discussed in detail in Chapters 2 and 3.

1.3 THE WIND HAZARD ON ROADS AND MOTORWAYS

Wind is not a major cause of severe road accidents. Price (1971) suggested that the total cost of wind as a hazard to traffic was of the order of £300 000 (equal to about £2½ million at 1991 prices), or some three orders of magnitude less than the total cost of road accidents. It is not significant on the national scale, but it can cause real problems for high-sided vehicles in exposed localities, and it is an important consideration in the design of bridges and the siting of road signs.

At the interface between the atmosphere and the surface of the earth, friction reduces mean wind speeds (an effect which can extend up to a height of 1 km in some areas) and makes the air turbulent. This turbulence shows itself in sharp fluctuations in wind speed (gusts and lulls) and changes in wind direction. Such gusting and turbulence, together with sharp velocity transition zones (where a vehicle emerges from a tunnel on to an exposed bridge for example), pose the greatest problems for the stability of vehicles, particularly high-sided lorries, double-decker buses, caravans and two-wheeled vehicles.

The force exerted by wind on a vehicle is proportional to the square of the wind speed and to the area of the vehicle presented to the wind (TRRL 1975); it is a maximum when the vehicle is broadside to the wind. This force is constantly changing due to the turbulent nature of the low-level airflow and the turbulent eddies induced by the traffic itself. If the vehicle is in motion any consideration of its stability becomes a complex dynamic problem involving not only the sideways overturning moment, but also the oscillatory forces at the rear of the vehicle and the mechanics of steering (Pritchard 1985). High-sided vehicles, and caravans or trailers being towed, may become unstable under strong gusts and may be blown over. The sudden gusts induced by the moving traffic may exacerbate the situation (Telionis *et al*. 1984).

The rider of a motorized two-wheeled vehicle is at a much greater disadvantage on the road or motorway in conditions of strong wind than drivers of other vehicles. The reasons for this are as follows.

(a) The rider of a motorcycle is likely to lose control of his machine more quickly than he would if driving a car, simply because of the loss of balance resulting from skidding, braking or swerving in high winds.
(b) Motorcycles are smaller than four-wheeled vehicles, and other drivers may not always see them in time, either because of lack of concentration or because the bike may be in a blind spot in the driver's mirrors–a particular problem if the motorcycle swerves across the road in conditions of strong winds.

(c) Motorcycle riders may be young and relatively inexperienced in reacting to an unexpected and potentially dangerous situation.

The Transport and Road Research Laboratory (TRRL 1975) has suggested that the wind speed which can cause accidents to vehicles is of the order of 15 m/s (33.5 mile/h), as measured by conventional anemometers at a height of 10 m above ground level, and averaged over 10 min, with gusts up to 22 m/s (49 mile/h). The frequency of occurrence of such gusts is fortunately not high over the British Isles. The average number of hours per year when the wind speed is equal to or exceeds 15 m/s (with gusts up to the above figures) is about four at Heathrow Airport and 39 for Prestwick in Scotland.

As an example of the problems presented by strong winds, on 1 November 1965 some 15 accidents were attributable to wind on a single section of the Al trunk road in Yorkshire. The highest mean wind speed recorded in the area that day was 19.5 m/s (43.6 mile/h). During the period 1966–68, 37 vehicles are known to have been overturned by the wind on the roads of Yorkshire; 20 more were blown off course sufficiently for them to strike safety fences, and 70 two-wheeled vehicles were involved in accidents where wind forces were believed to have played a contributory role (Price 1971).

The severe gale on 26 January 1990 illustrated the disruption that can accompany a major windstorm in the UK if it occurs during the working day. Deaths numbered 47 in England and Wales, many of them the result of vehicles running into fallen trees. Most motorways in southern Britain were closed for some hours, mainly because of high-sided vehicles being blown over.

There are no readily available national figures for the increase in road accidents at times of strong winds, mainly because this is not one of the weather types to be recorded on the National Police Accident Report Form instigated in 1969. Furthermore, non-injury accidents involving vehicles simply being blown over, do not have to be reported to the police. It is clear, however, that although the number of personal injury accidents from this cause is probably small, strong winds can be especially hazardous when they are present in combination with heavy rain, snow or ice, where vehicle control is less easy.

It is true that vehicles are at greater risk on more exposed stretches of road. Certain topographic configurations are known to present hazards of strong and turbulent wind when the wind blows from a particular direction:

(a) high-level routes generally experience stronger winds than low-level routes;
(b) gusts may be a particular problem where a road emerges from

a cutting on to a bridge or embankment section (particularly hazardous for caravans being towed in side-winds); and
(c) strong winds may present problems on bridges spanning deep valleys and cuttings, as there may be a pronounced wind-funnelling effect when winds blow along the line of the valley.

Even slight peculiarities in the local topography, however, may affect the wind field and traffic safety. For example, work near the Yoneyama Bridge in Japan, together with wind-tunnel tests (Narita and Katsuragi 1981) has shown how even a slight modification of a local topographic feature (the removal of a hump which caused locally dangerous wind conditions) can improve the safety of local driving conditions. The prediction of problem areas along a proposed route is not an exact science, however.

1.3.1 The assessment of exposure to strong winds

Exposure in a given locality is a complex function of the local topography and the prevailing wind field (Baker 1984). Wind exposure is difficult to assess at any particular site for, as Shellard (1965) of the Meteorological Office has stated: 'Making suitable allowances for topographical effects when estimating probable maximum wind speeds is not an easy matter; it usually consists of applying approximate rules or surmises which are based on general meteorological experience rather than on directly applicable observational evidence'.

Such observational evidence is best obtained from properly sited anemographs, providing a continuous record of wind speed and direction at a height of 10 m above ground level. Such equipment is expensive and bulky, and it is usually neither economic nor feasible to install large numbers of anemographs in a study area to obtain information on relative exposure. Four such anemographs were used by the author to monitor wind conditions in pre-selected locations between Dover and Folkestone in connection with a proposed new road scheme. Detailed information was obtained on the local wind climatology, but the equipment needs to be well protected from vandalism and maintained on a regular basis.

An alternative method which can be used to identify areas where exposure to wind is likely to be a problem is the use of wind-tunnel tests on scale models of the proposed motorway (Plate 1982). Such tests can be expensive, but useful results can emerge. Wind-tunnel tests were carried out in the design of the M62 motorway across the Pennines where it crossed the Scammonden Dam in Yorkshire; at this point the motorway is some 15 m above the reservoir surface, across which strong southwesterly winds can blow without obstruction for about 1 km. It was established that on the windward hard shoulder of the motorway, wind speeds could be up to 60% greater than over the surface of the reservoir. Substantial reductions were achieved with the provision of a wind-break about 1.4 m high and with a density of 40%; the wind-break has been designed to minimize its visual intrusion into the landscape (Price 1971).

A cheap, simple but effective method of obtaining information on the variability of exposure at a number of sites along proposed lines of new roads or motorways is the use of tatter flags. These are rectangular pieces of cotton (Madapolam, prepared to a British Standard), 38 cm by 30 cm in size. The flags are mounted in a standard manner on metal rods (over 1 m long) to rotate freely. Field tests and wind-tunnel tests have shown that the amount of area loss or tatter of these flags is directly related to the square of the mean wind speed at the site (Rutter 1968) and gives a realistic assessment of relative exposure. A large number of flags can be installed at field sites of interest, collected after a suitable period of time, and replaced. Tatter flags are cheap, easily produced and require no maintenance. The major sources of error arise from interference by man or beast. An attempt by the author to use them at 15 sites in the Dover–Folkestone area to provide a survey of exposure and local wind hazards (in conjunction with the four standard anemographs described previously) along alternative lines for a proposed new road scheme was thwarted by local vandalism. Cows and sheep have also been known to chew the flags.

The tatter flag was pioneered and developed by the Forestry Commission, which uses it extensively to define tree-planting limits in upland regions (Lines and Howell 1963). British Rail has used tatter flags to investigate exposure along high-level stretches of track used by its high-speed trains. McAdam (1980) used the method successfully in the Falkland Islands to assess variations in exposure, while the Royal Society has used tatter flags in the Andaman Islands to study shelter effects.

1.3.2 Reduction of the wind hazard

To alleviate the problems of strong winds on roads and motorways in exposed locations, a number of preventive and remedial measures may be taken. The preventive measures will alter the prevailing wind field to reduce mean wind speeds, and will include the careful design of road sections to avoid sudden transitions from cutting to embankment sections, and the use of wind-breaks.

Wind-breaks such as that used on the M62 can take a variety of forms, varying from natural tree screens to solid parapets on bridge decks.

Anything which protrudes above the general road level will in some way influence the wind effect on road traffic. Natural trees may not grow to their expected heights in exposed coastal locations, however, due to the combined effects of the strong winds and the high salt content of the air.

Artificial wind-breaks have been used at a number of sites in Britain and have usually taken the form of a semi-permeable wooden fence (i.e. a fence with gaps in it). The Transport and Road Research Laboratory has carried out work which suggests, as a general guide, that a 50% reduction in wind speed across a road can be obtained either with a 60% density windbreak (60 mm slats, 40 mm gaps) 3 m high, or an 80% density windbreak 2.5 m high positioned near the verge (Hay 1971).

Wind-breaks can have undesirable side-effects, and can generally only be used after field tests have been carried out. Some of the problems are as follows.

(a) In certain wind directions, the wind may be funnelled between the wind-breaks and may therefore be increased; the windbreak is therefore normally positioned at right angles to the prevailing wind where feasible.
(b) Wind-breaks may cause undesirable snow drifting in winter.
(c) Road traffic emerging from a wind-sheltered area behind a wind-break may be suddenly exposed to side winds.
(d) Wind-breaks 3 m high may not be very attractive visually and may increase the wind forces on the superstructure of a long-span bridge if used in such situations.
(e) There is a slightly increased risk of ice on the road in winter where the windbreak screens the road surface from direct sunlight. Nevertheless wind-breaks have been very successful when used in the right circumstances and they can reduce potentially dangerous wind speeds to those which are within acceptable limits. They have an indirect advantage in the local environment in reducing motorway noise.

The remedial measures which can be taken in particular problem locations include the following.

(a) Speed restrictions

Traffic on the Severn Bridge is subject to prescribed traffic management and operational guidelines (Dept of Transport 1986). When wind-speeds reach approximately 35 mile/h (15 m/s), consideration is given to the imposition of speed limits and the restriction of traffic to a single lane on each carriageway (depending on wind direction and other circumstances). At windspeeds of 40 mile/h (17.5 m/s), one lane is closed to traffic regardless of circumstances. When windspeeds on the crossing reach approximately 45 mile/h (20 m/s), the police close the Bridge to high-sided vehicles; the crossing continues to operate with a single lane open on each carriageway. When windspeeds exceed 55 mile/h (25 m/s) the police obtain a forecast windspeed from the Bristol Weather Centre, and if this exceeds 62 mile/h (27.5 m/s) the Bridge is closed to all traffic; in the absence of a forecast of winds exceeding 62 mile/h traffic restrictions are maintained. If observed winds on the bridge exceed 80 mile/h (35 m/s) the bridge is closed to all traffic immediately.

(b) Warning signs

On the Forth road bridge in Scotland, illuminated 'strong wind' signs are turned on if the wind exceeds 30 mile/h (13.4 m/s); warning signs are also used on the approach to the exposed Thelwall Viaduct on the M6 motorway in Cheshire.

(c) Radio and television broadcasts

Bad weather warnings may be 'flashed' to advise drivers to avoid particular problem areas. An enquiry into the damage caused by severe winds in Britain led to revision of the form of flash warnings by the Meteorological Office as follows:

Table 1.2 *Flash warnings in UK: criteria for severe gales*

Tier 1	Mean speed	20.5 m/s
	Gusts	30.8 m/s
Tier 2	Mean speed	17.4 m/s
	Gusts	22.1 m/s

Source: Hunt (1990)

(d) Road closure

As noted above this operates on the Severn Bridge during periods with winds of over 62 mile/h. There is no national policy to control vehicle movement on bridges and exposed stretches of road during strong winds. Baker (1987) has advocated a simple two-level system of control: (i) when the wind gusts above 40 mile/h (17.5 m/s), a speed limit of 22 mile/h (10 m/s) should be placed on all high-sided vehicles, and appropriate warning signs should be activated; and (ii) when the wind gusts above 50 mile/h (22.5 m/s) all vehicle movement should be stopped.

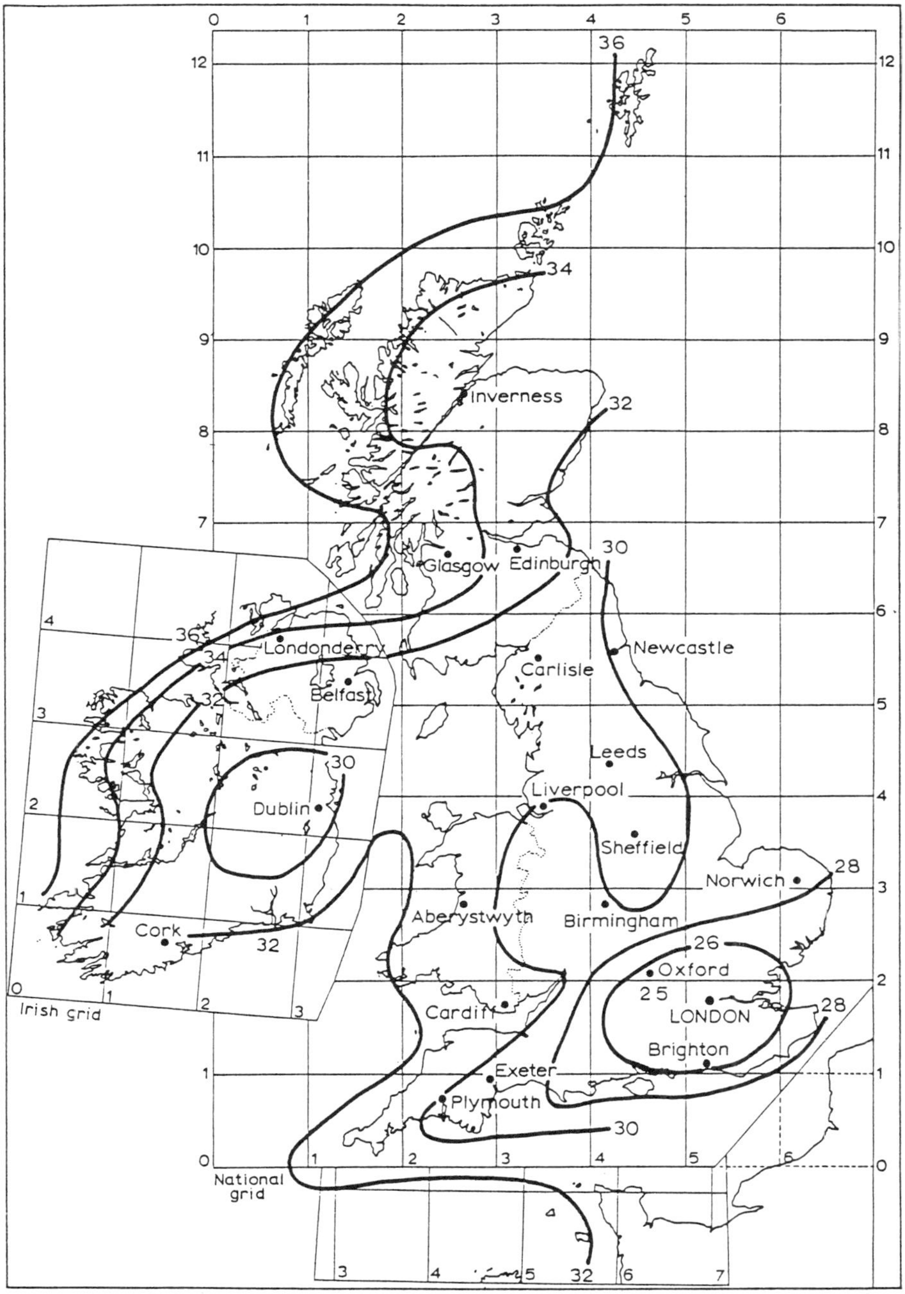

Figure 1.2 Mean hourly wind speed (in m/s) likely to be exceeded once in 120 years at 10 m above the ground in level, open country. Isopleths are drawn at 2 m/s intervals (Meteorological Office, from Hay 1974).

1.4 WIND AND BRIDGES

All structures such as bridges resist the flow of air, so that the wind correspondingly exerts a force on the bridge. There are two types of force or loading involved on a bridge structure: *static wind loading* on the exposed area of the bridge, giving the bridge a slight but steady deflection; and *dynamic wind loading*, producing motion or oscillations in the structure arising from interactions of the bridge itself with the airstream (aerodynamic instability). The latter is particularly important in the design of suspension bridges. Bridges have been damaged or destroyed by both the static and the dynamic actions of wind. The static wind loading was largely responsible for the collapse of the Tay Railway Bridge in 1879, while dynamic forces caused the destruction of the Tacoma Narrows Bridge in 1940, when a vertical bending oscillation in the structure changed to a torsional oscillation which shook the bridge to destruction in less than one hour, in winds of only 42 mile/h (18.7 m/s). Such self-induced oscillations may now be largely prevented with careful design (National Physical Laboratory 1955) and engineers have become experts in handling such design problems.

The problems of estimating wind speeds and wind loadings for the design of bridges are well discussed in Kerensky (1971) and Hay (1974). A bridge designer needs to know the maximum wind speed to be expected at the bridge during the lifetime of the structure. Bridge design specification today is based on the mean hourly wind likely to be exceeded once in 120 years at a height of 10 m above level ground. The spatial variation of this over the British Isles is shown in Figure 1.2. This datum speed is then adjusted to allow for the degree of wind funnelling (if the bridge is located in a deep valley or where lee-wave effects occur), the height of the structure, local topography, and the anticipated lifetime of the bridge.

However, as Price (1971) has stated: 'Modern designs of large suspension bridges, which are intended to minimize the wind force on the structure, do not necessarily at the same time also reduce the forces on the vehicles using the bridge'. This is an important point which has clear implications for the safety of road traffic. It is essential that wind-induced accidents are minimized on bridges because of the potential seriousness of the consequences for the bridge, the vehicles involved and their occupants. Problems still exist for vehicles emerging from sheltered cuttings on to exposed bridges with strong side winds. Indeed, measurements suggest that the mean wind speed at the centre of a bridge may well be double that at the sides, with maximum gusts about $1\frac{1}{2}$ times those at the sides. The problem may be ameliorated by suitable landscaping and design of the crossing, together with appropriate warning signs on the road and other controls.

1.4.1 The Severn Bridge

The Severn Bridge, carrying the M4 motorway over the River Severn, is known to be vulnerable to adverse weather conditions, particularly strong winds, and the operational guidelines used to control traffic movement in these situations have already been outlined. Wright (1985) has described some of the techniques used by the Meteorological Office to predict strong winds, gusting to 35 mile/h or more over the Severn Bridge.

Traffic records on lane closure on the Severn Bridge (Department of Transport 1986) indicate that:

(a) traffic crossing the Bridge is affected by lane closures due to high winds on an average of some 130 hours per annum on 20 days per year;
(b) the lane closures occur almost entirely during the autumn and winter months, December and January accounting for 65% of the times of closure;
(c) the average duration of lane closure is about seven hours, although some closures have exceeded 20 hours, and nearly half are three hours or less; and
(d) three-quarters of the closures occur between 0600 and 2200 h.

The Severn Bridge is affected by winds strong enough to close the crossing to high-sided vehicles for some 20–25 hours annually. Total closure of the bridge to all vehicles is an extremely rare occurrence.

The relationship between critical mean wind speeds at the bridge deck height and the number of hours per annum when such speeds are exceeded is shown in Figure 1.3. The relationship is a sensitive one as shown in Table 1.2.

A second Severn Bridge road crossing is currently being planned, and is due for completion by around 1996. Consultants consider that the provision of a combination of barrier-type aerodynamic fairings, a suitable bridge-deck cross-section and structural stiffness would partially shield traffic from the effects of strong winds and would considerably reduce the need for lane closures and other traffic restrictions.

1.4.2 The tornado risk at bridges: the example of the Severn Bridge

Bridges are not immune from the impact of tornadoes and their accompanying violent winds. Elsom (1985) has shown that since 1971 Britain has experienced an average of some 18 tornado days per year, with an annual average of 45 separate tornadoes reported; in 1981 there were 152 reported tornadoes on only 12 days. The tornado hazard is often underestimated or neglected when considering design criteria.

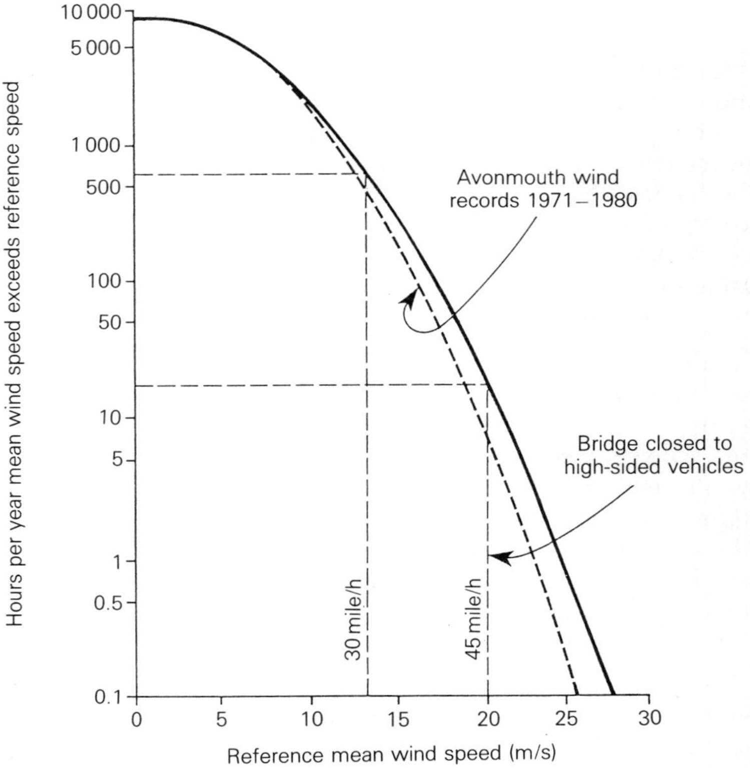

Figure 1.3 The relationship between mean wind speeds on the Severn Bridge and the average number of hours per annum when such speeds are exceeded (from Department of Transport 1986).

Table 1.3

Critical mean wind speed on the bridge deck (*mile/h*)	(*km/h*)	*Number of hours when critical wind speed is exceeded per annum*
30	48	600
35	56	220
40	64	75
45	72	20

On 1 March 1981 at 1645 GMT a tornado column passed straight through the western end of the Severn Bridge. Fortunately it just missed vehicles which were on the bridge at the time, but it destroyed a caravan and a garage nearby, destroyed trees in its path and removed tiles from houses in the locality as it passed through (Meaden 1983). The width of the damaging winds has been estimated at 40–45 m; the tornado had a land-track of 1 km and a water track (as a waterspout) of 6–8 km.

It was indeed fortunate that there were no vehicles in its path as it crossed the bridge; what might have happened to them was clearly demonstrated only 200 m further on where the edge of the tornado completely destroyed a caravan. The strength of the wind, estimated at 33–41 m/s or 73–92 mile/h, was well shown by the fact that a large part of the roof of a building in Beachley, only 150 m south of the bridge, was torn off and hurled at the bridge supports.

Meaden (1985) emphasizes that, in each decade, Britain experiences hundreds of known damaging tornadoes, a sizeable fraction of which have winds strong enough to put the wind-design basis of such a bridge to the test. Damage to vehicles caught on a bridge during the passage of a tornado would normally be severe. He has estimated from available data that the return period for any tornado over the 2 km-long Severn Bridge is of the order of 40 years. Such a figure is only an average time interval. It does not mean that it will be 40 years before another tornado crosses the bridge: it may be more, but it might be less. Such tornadoes have a range of intensities, however, and some 40% of British tornadoes are more violent than that which passed across the bridge in 1981. For example, in October 1913, a devastating tornado with winds of 84–95 m/s (up to 200 mile/h) occurred in South Wales, only 40 km west of the Bridge.

When data on the risk from tornadic winds are combined with wind estimates produced from conventional anemograph records, then the design gust-return periods are reduced and the maximum gust speeds likely to occur within stated time intervals are increased.

1.5 WEATHER-RELATED PROBLEMS AND ROAD CONSTRUCTION

1.5.1 Soil moisture

Most landslips and slumps in roadside cuttings occur in structurally weak soils or sediments, where the side slopes have exceeded their angle of local equilibrium and the slope fails. A landslip across a motorway cutting during the hours of darkness can be very dangerous. Landslips are essentially a problem for the engineer rather than the meteorologist, but they are nearly always triggered by changes in the soil moisture content following heavy rainfall; they thus have a hydrological/meteorological component.

Sherrell (1971) has emphasized the need for detailed knowledge of the characteristics of the local aquifer and groundwater if landslips such as those which have occurred on the Cullompton bypass in Devon are to be avoided. From their experience of the instability of cuttings on the M4 north of Cardiff, Newberry and Baker (1981) have re-emphasized this plea and stress the need for continuous geotechnical monitoring during the construction of all major cuts in soft rocks. If landslipping is considered to be a potential hazard, likely to be triggered by the combination of heavy rainfall on soft, weak sediments with a high soil moisture content, then massive berms or ledges can be employed to stabilize the mass at risk, as used on the A40 near Monmouth (Early and Jordan 1985).

1.5.2 Frost heave

Frost heave occurs beneath roads when soil moisture in the base course beneath the asphalt freezes. This only happens during long spells of severe cold weather, for rarely do soil temperatures at depth fall below freezing in Britain. Freezing causes an expansion of the water content by 9% of its original volume, and frost heave causes a loss of cohesion and strength in the road suface, which is then easily broken up (Jacobs and West 1966). At any one location, frost heave occurrence is a function of the prevailing temperatures, the position of the water table beneath the road, the properties of the materials used and the depth of frost penetration. If the risk is to be avoided in known frost hollows, then there is a need to keep water out of the road structure, with the use of drains to maintain the water table at depths of one metre or more.

1.5.3 Effect of weather on road construction work

Weather conditions during the period of road or motorway construction are economically important, for delays may escalate the cost of construction considerably (Washington Transportation Board 1978). Before bidding for a road construction contract, firms need good climatological information on the number of days on which work would normally be possible in order to determine completion dates. Maunder (1971, 1986) has developed a simulation model incorporating a soil moisture index, to estimate the effects of climatic variables on road construction. A similar study by Attanasi *et al.* (1973) has examined the impact of soil moisture, precipitation and daily temperature. Models were developed to translate observed weather conditions into probabilities for carrying out particular construction activities, namely paving, asphalt work and the construction of bridges and drainage structures. The estimates produced indicate a strong sensitivity to local precipitation and soil moisture conditions.

Although no precise limits can be given, the Meteorological Office and the Building Research Station have suggested that construction is likely to be interrupted whenever temperatures fall below 1 °C, and the rate of rainfall exceeds 0.5 mm/h.

1.6 REFERENCES

Attanasi, E.D. *et al.* (1973). Forecasting work conditions for road construction activities: an application of alternative probability models. *Monthly Weather Review*, **101**, 223–230.

Baker, C.J. (1984). Determination of topographic exposure factors in complicated hilly terrain. *Journal of Wind Engineering & Industrial Aerodynamics*, **17**, 239–249.

Baker, C.J. (1987). Measures to control vehicle movement at exposed sites during windy periods. *Journal of Wind Engineering & Industrial Aerodynamics*, **25**, 151–62.

Chaplain, R. (1987). Weather Watchers-the new way to provide weather services. *Journal of Meteorology*, **12**, 233–239.

Codling, P.J. (1974). Weather and road accidents. In *Climatic Resources and Economic Activity*, ed. J.A. Taylor. Newton Abbott: David & Charles.

Department of Transport (1986). *Second Severn Crossing Report*. London: HMSO.

Early, K.R. and Jordan, P.G. (1985). Some landslipping encountered during construction of the A40 near Monmouth. *Quarterly Journal of Engineering Geology*, **18**, 207–24.

Elsom, D.M. (1985). Tornadoes in Britain: when, where and how often? *Journal of Meteorology*, **10**, 203–211.

Hay, J.S. (1971). Reduction of wind effects on vehicles by the use of wind breaks. *Transport and Road Research Laboratory Technical Note 690*. Crowthorne: TRRL.

Hay, J. S. (1974). Estimation of wind speed and air temperature for the design of bridges. *Transport and Road Research Laboratory Report LR 599*. Crowthorne: TRRL.

Hogg, W.H. (1973). *Report on Fog with Particular Reference to the M40 Oxford-Birmingham Motorway, Warwick Section* (Midland Road Construction Unit, Department of Transport). 55pp.

Hunt, R. (1987). Better forecasts aid war against winter weather. *New Civil Engineer*, **29**, 32.

Hunt, R.D. (1990). Disaster alleviation in the United Kingdom and overseas. *Weather*, **45**, 133–8.

Jacobs, J.C. and West, G. (1966). Investigations into the effect of freezing on a typical road structure. *Transport and Road Research Laboratory Report No 54*. pp 49, Crowthorne: TRRL.

Kerensky, O.A. (1971). Bridges and other large structures. *Philosophical Transactions Royal Society London (A)*, **269**, 343–352.

Lines, R. and Howell, R.S. (1963). The use of flags to estimate the relative exposure of trial plantations. *Forest Record No 51*. Forestry Commission.

Lovell, S.M. (1966). Design of roads to minimise snow, fog and ice problems. *Municipal Engineering*, **143**, 1369–1373.

Maunder, W.J. *et al.* (1971). Study of the effect of weather on road construction: a simulation model. *Monthly Weather Review*, **99**, 939–945.

Maunder, W.J. (1986). *The Uncertainty Business*. London: Methuen.

McAdam, J.H. (1980). Tatter flags and the climate in the Falkland Islands. *Weather*, **35**, 321–327.

Meaden, G.T. (1983). The Severn Bridge tornado-waterspout of 1 March 1981. *Journal of Meteorology*, **8**, 37–45.

Meaden, G.T. (1985). Tornado-waterspout risk at the Severn Bridge. *Journal of Meteorology*, **10**, 239–242.

Meteorological Office (1985). *The Susceptibility of Fog on the M25 Motorway*, Bracknell: Building & Construction Climatology Unit, Meteorological Office.

Musk, L.F. (1982). The local fog hazard as a factor in planning new roads and motorways. *Environmental Education & Information*, **2**, 119–129.

Narita, N. and Katsuragi, M. (1981). Gusty wind effects on driving safety of road vehicles. *Journal of Wind Engineering & Industrial Aerodynamics*, **9**, 181–191.

National Physical Laboratory (1955). *Wind Effects on Bridges and other Flexible Structures*. London: HMSO.

Newberry, J. and Baker, D.A.B. (1981). The stability of cuts on the M4 north of Cardiff. *Quarterly Journal of Engineering Geology*, **14**, 195–207.

OECD (1976). *Adverse Weather, Reduced Visibility and Road Safety*. Paris: OECD.

Perry, A.H. (1981). Snow, frost and cold. In *Environmental Hazards in the British Isles*, pp. 43–69. London: George Allen & Unwin.

Perry, A.H. and Symons, L. (1980). Economic and social disruption arising from the snowfall hazard in Scotland – the example of January 1978. *Scottish Geographical Magazine*, **96**, 20–25.

Pike, W.S. (1986). The intersection of a sudden hailstorm with Motorway M4 Junction 14 in West Berkshire, 0850h on 9 November 1985. *Journal of Meteorology*, **11**, 51–55.

Plate, E.J. (ed.) (1982). *Engineering Meteorology*. Amsterdam: Elsevier.

Price, B.T. (1971). Airflow problems related to surface transport systems. *Philosophical Transactions Royal Society London A*, **269**, 327–333.

Pritchard, R.J. (1985). Wind effects on high-sided vehicles. *Journal of Institute of Highway Transportation*, **56**, 22–25.

Robinson, A.H.O. (1965). Road weather alerts. In *What is Weather Worth?*, pp. 41–43. Melbourne: Australian Bureau of Meteorology.

Rutter, N. (1968). Tattering of flags at different sites in relation to wind and weather. *Agricultural Meteorology*, **5**, 163–181.

Shellard, H.C. (1965). The estimation of design wind speeds. In *Wind Effects on Buildings and Structures*, National Physical Laboratory Symposium No 16, pp. 29–51. London: HMSO.

Sherrell, F.W. (1971). The Nags Head landslips, Cullompton Bypass, Devon. *Quarterly Journal of Engineering Geology*, **4**, 37–73.

Sherretz, L.A. and Farhar, B.C. (1978). An analysis of the relationship between rainfall and the occurrence of traffic accidents. *Journal of Applied Meteorology*, **17**, 711–715.

Smith, K. (1982). How seasonal and weather conditions influence road accidents in Glasgow. *Scottish Geographical Magazine*, **98**, 103–114.

Telionis, D.P. *et al.* (1984). An experimental study of highway aerodynamic interferences. *Journal of Wind Engineering & Industrial Aerodynamics*, **17**, 267–294.

Transport and Road Research Laboratory (1975). Wind forces on vehicles. *TRRL Leaflet 445*, Crowthorne: TRRL.

Washington Transportation Board (1978). *Effect of Weather on Highway Construction*. Washington, 29 pp.

Wright, K.C. (1985). Techniques for forecasting the occurrence of strong winds over the Severn Bridge. *Meteorological Magazine*, **114**, 78–85.

Chapter Two

The winter maintenance of highways

A.H. Perry and L.J. Symons

Ice and snow on the highway reduce friction between the tyre and the road surface. In areas where these phenomena are frequent and persistent the problem is often overcome by a combination of action by the road user (e.g. fitting special snow tyres), and preventive measures by the highway authorities (e.g. clearing snow or treating roads to prevent ice formation). It is in areas where the hazards are intermittent or occasional that the impact on traffic movement is often greatest. In such areas the winter maintenance of highways often involves considerable expenditure on manpower, equipment and resources in order to keep highways in a condition that is safe for road users. Typically this involves close co-operation between meteorologists, who are called upon to provide advance warning of dangerous conditions, and highway engineers who must be available to take remedial action.

The United Kingdom offers a good example of a country where over a long period of time a partnership has been established between meteorologists and highway engineers to tackle a winter problem that is highly variable in its impact from year to year. With average daily minimum temperatures hovering close to freezing point in most places for at least three months of the year, on any individual night there is a high probability, but not a certainty, that ice could form if roads remain untreated.

2.1 SALT APPLICATION

Although it is generally recognized that it is difficult to find a practical alternative to the application of salt to reduce the risk of motor vehicles skidding on roads affected by ice, there are serious economic and environmental objections to its excessive use, as well as the incidental damage it causes to vehicles and other metal structures such as bridges. As the costs of a salting run and of alternative chemicals such as urea and glycol are known to each highway authority, there is no need to comment here on the economic aspects of the matter. In the UK, there has been little attention to the environmental aspects, which have caused considerable arguments on the continent of Europe. In both Germany and Switzerland, where there is considerable concern at damage to forests, and motor vehicle exhaust fumes are widely believed to be one of the causes, excessive salt applications are also regarded with suspicion. In Berlin, salt in the groundwater is held to be one of the dangers to trees, of which about 50 000 show leaf damage partly attributed to salt.

Despite environmental objections there appears to have been little attempt to reduce salt application to roads in either Germany or Switzerland. It is generally considered that the risk of accident is so much greater on roads not treated with salt in winter that the public would prefer the economic and environmental consequences of continuing to salt. In Berlin, however, there has been a major change towards minimizing salt applications even though the new measures cause serious inconvenience to all road users. In 1983, a law was introduced with the agreement of all political parties to eliminate salt application except in selected locations. It is also legally required that applications should not exceed $40\,g/m^2$ in spite of the very severe winter climate. Reliance is now placed on applications of various types of grit, though it is necessary to use much more grit than salt in the same conditions; and since the grit quickly loses effectiveness, it is often necessary to make gritting runs three times in 24 hours, possibly even eight to ten times more frequently than with salt. The demand for labour and vehicles is clearly very high but the favourable Berlin financial situation minimizes this drawback and it is recognized that such a procedure does help to combat any rise in unemployment.

Other drawbacks to the system include the accumulation of large quantities of grit beside the roads, necessitating a month-long clearance operation in spring, even with additional casual labour. Piles of grit, and icy conditions at bus stops, make it more difficult for buses to pull in, and increase the frequency of accidents. Organized cycling interests have requested that the grit should not be spread on cycle paths, preferring the risk of skids to the frequency of punctures experienced when the new system was introduced. Car tyres also suffer and there has been an increase in the number of eye complaints reported.

In the UK because of both the cost of salt and its unwelcome environmental impacts, the highway engineer is interested in using the minimum amount of salt necessary to maintain the highway in a safe condition, while at the same time ensuring that the statutory requirement to keep the highway safe is met. This has been achieved for many years by purchasing a specialized weather forecast from the Meteorological Office. This Road Danger Warning Service was generated on a national basis and disseminated daily using a standard format. By the mid 1980s there was considerable dissatisfaction amongst highway authorities, particularly concerning the level of accuracy that was being provided, with evidence that as many as 30–40% of forecasts were too pessimistic (Ponting 1984).

In an effort to improve the service that was being offered, and extend the amount of information on which the decision would be taken on whether or not to treat the roads, the service was re-organized and re-launched under the title 'Open Road'. As well as giving such detail as site-specific snow advice prediction and a five-day planning forecast, Open Road includes products that are the result of new technology. These include annotated hard copies of digitized weather-radar images.

2.2 WEATHER RADAR AND THE HIGHWAY ENGINEER

The UK weather-radar network was declared operational at the beginning of 1985 and now represents the most extensive and technologically advanced system of its kind in Europe. Experimental work had been in progress for several years, with a mobile radar installation in North Wales in the early 1970s. By 1980, four weather-radar installations were producing precipitation data in real time and, since then, two more have been added to the network in the UK and one has been installed in Eire. It is envisaged that eventually a network of radars will provide a coverage of the entire country (Figure 2.1).

From the outset, weather radar in England and Wales has been seen primarily as an aid to flood forecasting, and groupings of water authorities, the Meteorological Office and the Ministry of Agriculture have been involved in providing funding for radar installations. In Scotland, progress in the installation of weather radar has been slower than in the rest of the UK, since the potential benefits for flood alleviation are perceived as smaller although, significantly, road transport has been identified in a recent report as potentially benefiting.

Weather radar identifies and quantifies areas of precipitation detected within the radar field. Maximum radar range is 210 km, although frequently precipitation is not detected beyond the 100 km limit. A central network computer at the Meteorological Office combines data from all the radars into one 'composite' image which is updated every 15 min

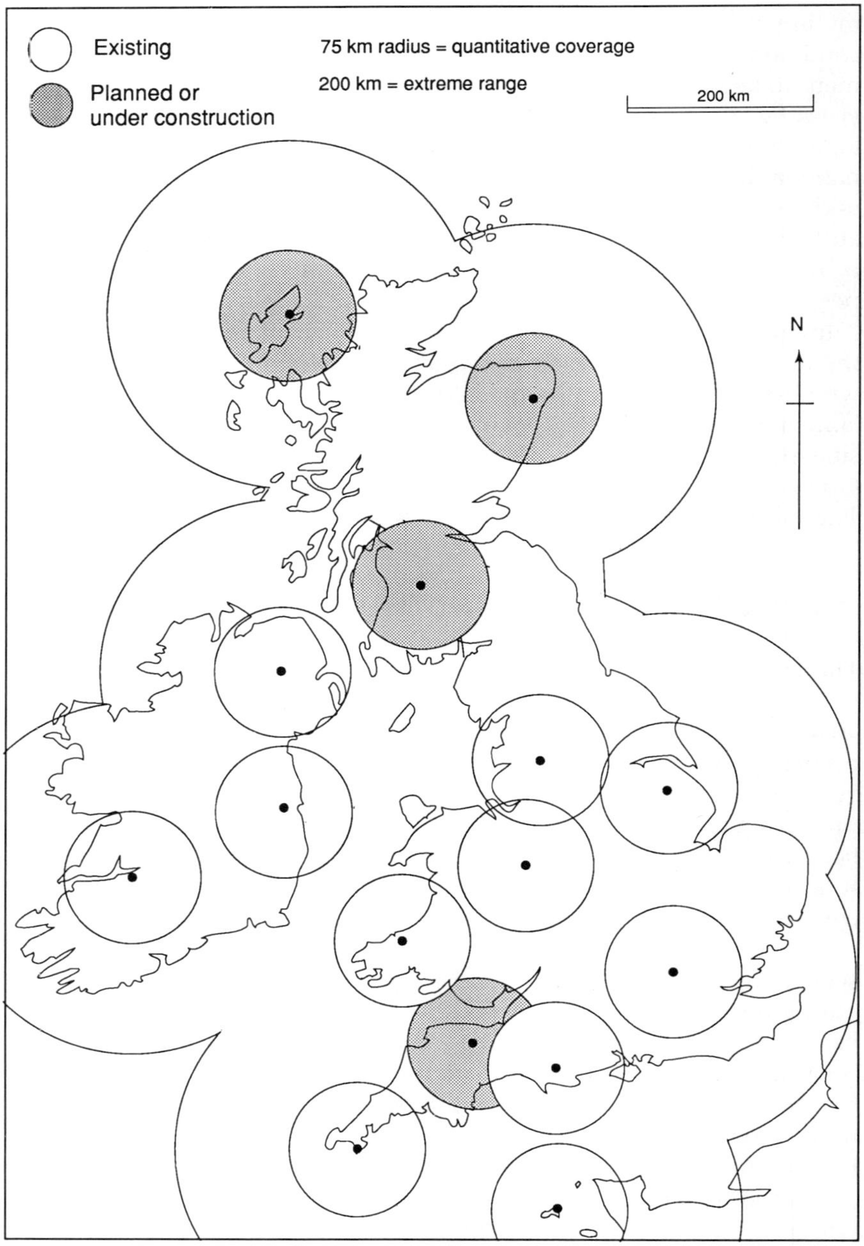

Figure 2.1 Weather radar sites operational, under construction and planned in the British Isles in 1992.

and transmitted to the user via dedicated British Telecom lines. The visual display of the current pattern of precipitation over the UK employs eight stepped intensities, each represented by a different colour. The colour-coded pixels making up the composite image give a resolution equal to a 5 km square. Precipitation type is not readily discernible from the weather-radar display system, and various sources of error, such as spurious generation of apparent precipitation, may contribute to an incorrect estimate of probable rainfall; it is therefore advisable to use the radar in conjunction with a consultancy service provided by a Weather Centre. To monitor the progress of precipitation patterns through time, it is important to be able to store and retrieve historic radar images, and this facility has been developed and recently much refined. At any time stored images can be replayed in sequence, allowing the user to monitor the development and movement of specific precipitation configurations. There is also the potential to incorporate a floppy-disk unit in the storage hardware, which considerably enhances the total data-storage capability.

While the possibility of making radar imagery widely available through teletext and viewdata services has been discussed, this seems unlikely to take place in the foreseeable future in the UK, and there are no plans to offer a weather radar service in this form. Consequently, if the highway engineer wishes to make use of weather radar, the only option is to purchase the weather-radar imagery and the microcomputer equipment required to display it. Costs are high, since the investment in providing the system is large and the major customers, the water authorities, have shown that the market will tolerate a high price for the service. The radar facility's ability to pay for itself, in terms of winter maintenance and other cost savings over specific periods, has not yet been fully demonstrated. Weather radar could help to reduce road traffic accidents by providing local information on precipitation which could be used to prevent the development of hazardous road surface conditions. Research at University College, Swansea has identified a number of situations where cost savings could be expected as a result of having direct access to weather-radar information. These include two common weather scenarios:

(a) On cold, showery winter nights, individual showers can be tracked. Wetted roads may freeze over once skies clear after showers, producing locally hazardous early morning road conditions. During such weather conditions, the showers which are generated over the warm seas around the UK drift across coasts and inland before dying out. Traditional forecasts cannot provide the fine detail of shower tracks and local authorities have little choice but to salt an entire road network, much of which may remain dry and thus not be subject to an ice hazard.

(b) Following heavy snowfall, information on the time of cessation of the snowfall and the likelihood of recurrence can optimize clearing operations and prevent clearance work being undone by new snowfalls. Traffic management, particularly on motorways, could be enhanced by the use of weather radar for the determination of advised speeds on illuminated matrix signals.

If the benefits of radar imagery are to be fully achieved, then working practices may have to be reconsidered. Clearly, 24-hour working by personnel trained in the interpretation of weather radar-imagery is essential and, as in the case of Devon, this suggests a centralized control room or county headquarters making decisions. This would be preferable to decentralized decision-making. There may be a case for groups of neighbouring counties' sharing the cost of such a facility. A further important issue is whether radar imagery can provide precipitation warnings quickly enough to ensure the effectiveness of operations carried out by users. The question of warning provision has been the subject of recent research at Swansea. Results suggest that composite weather-radar imagery could provide visual warnings of precipitation of up to three hours on about 85% of occasions and this should be sufficient to ensure the effectiveness of such winter maintenance operations as precautionary salting and the call-out of snow-ploughs (Perry *et al.* 1986).

2.3 WEATHER INFORMATION FOR THE DRIVER

For the travelling public, broadcast information is often the only means of providing information on expected hazards, including those such as snow which could lead to the cancellation or restriction of unnecessary journeys. It is also in the interest of highway authorities that there should be minimum traffic on roads at such times since broken-down vehicles can cause obstructions and hinder road-clearance operations.

As well as the normal schedule of weather forecasts on radio and television, traffic information bulletins also include mention of winter weather hazards on roads. However, many of these reports are too short, are variable in their content and frequently contain too little information for the intending road user easily to appraise conditions, particularly a driver contemplating a long cross-country journey. Expectations by the forecaster of severe weather, for example snow falling and accumulating at a rate of 2 cm/h or more, will trigger a

FLASH message which will be broadcast on radio and television, interrupting scheduled programmes if necessary. For such messages to be issued there must be a high degree of confidence that the forecast conditions will materialize since they are designed to have considerably more impact than routine forecasts or warnings.

Information on road conditions is also available from teletext services like BBC's Ceefax, from local radio stations, and from the motoring organizations. Gardner (1987) has suggested that the only way to improve information significantly is to input more accurate information into a national viewdata system which can be interrogated by road users for local area or specific route information. This implies that the road user is prepared to invest in a viewdata facility, which may not be the case with the private motorist, although business interests could quickly adapt such a system if it were cost-effective.

Considerable research activity is in progress on driver communication systems. Developments in fibre-optics technology have led to improvements in variable signalling panels of the type often supported on overhead gantries above busy roads. New driver communication systems, including dashboard displays using radio transmissions, or inductive loops in the road, should soon become more widely available to supply weather information to the road user.

2.4 TRAFFIC DISRUPTION FROM THE SNOW HAZARD IN THE UK

The degree of disruption caused by a particular snowfall is determined by two main sets of factors: (a) the physical nature of the fall, including snow depth and duration and the associated meteorological conditions such as wind speed which may cause drifting, and the nature of the snow; and (b) the state of preparedness of the community and the individual – this will be affected by the frequency with which heavy snowfalls are perceived to occur, and the investment of capital in snow-clearing operations.

It may seem surprising that major road dislocation can, from time to time, affect a country like the UK where local-authority road-clearing departments are relatively well equipped and organised. Perry and Symons (1980) have described the blizzards in northern and eastern Scotland in January 1978 which paralysed the region, trapping several hundred cars in deep snowdrifts and necessitating an airborne rescue operation to rescue trapped drivers. Six years later, in 1984, level snow depths of 50–60 cm, with deep drifts, again caused blocked roads and prolonged disruption of communications in the Highlands of Scotland. Snow gates have now been installed on several roads. One of the main advantages that they confer is that, if vehicles can be prevented from entering and becoming snowed-up on difficult stretches of road like the Drumochter Pass on the main A9 routeway, snow-clearing operations can be carried out unhampered, but there has been local community resistance to their installation.

Although heavy falls of snow are less frequent further south over England and Wales than in Scotland, their impact in densely populated areas can be even more disruptive. Table 2.1 shows the nature that disruption can take with different amounts of snow lying. While in the countryside snow can be blown from fields to fill narrow roads and lanes which act as natural receptacles, in towns, city streets may require not only clearing but also snow hauling to open spaces since the physical space to dump cleared snow is often not available (Perry 1981).

The paralysis which followed severe blizzards in Wales and South-West England in both 1978 and 1982 highlighted the importance of road communications even in the depth of the country where commercial dairy farming is dependent on regular, efficient transport of milk from the farms. When this becomes impossible thousands of gallons of milk may have to be poured away, leading to considerable financial loss. It is estimated that in 1990 there were over 700 000 heavy and more than two million light goods vehicles on Britain's roads. Even a few days' disruption to a significant part of this fleet results in the total bill to the economy running into millions of pounds. Not surprisingly it is cost-effective to invest heavily in plant to clear roads as quickly as possible, even though for the fast removal of deep snow, self-propelled snow blowers, costing over £100 000 each, are essential. Twenty-four-hour working on snow clearance of the Primary Route Network (about 15% of the total road mileage) will be started once severe conditions develop. Not surprisingly expenditure on winter maintenance can be a significant part of total highway expenditure: in the Highlands of Scotland it constitutes 25% of the total road budget. In January 1987, in Kent, the highway authority mobilized 10 000 items of plant and was spending £5 million a day (Manson, 1987). As Figure 2.2 shows, in a typical lowland England county like Essex, expenditure is highly variable from year to year, depending on individual winter severity.

Those responsible for the transport systems are not judged by what resources or effort they put in but what they achieve in respect of keeping transport systems operating. Local authorities in England and Wales spend around £100 million a year on salting and snow-clearing operations (Figure 2.3) with winter maintenance accounting for 8.5% of

Table 2.1 *A hierarchy of disruption due to snow and ice in the British Isles*

	Effects	*Comments*
1st order – paralysing	Blocked roads, paralysis of communications, isolated rural settlements, power failures, major absenteeism in schools and at places of work. Rescue work undertaken by air, major livestock losses on hill farms.	25–30 cm of level undrifted snow. Return period in southern England about 15–20 years.
2nd order – crippling	Communications difficult with only essential journeys by road advised by police. Cancellation of most outdoor sporting events. Closure of rural schools.	Accompanies 10–20 cm of level snow in lowland England. Return period of 5–10 years in the south of England.
3rd order – inconvenience	Traffic movement impaired with increase in accidents attributable to snow and ice conditions. Sporting events affected.	Occurs on average every other year in southern districts.
4th order – nuisance	Traffic movement slowed, minimal media coverage. Gritting lorries rather than snowploughs required.	Can be expected to occur every winter, except in extreme SW

the total highways maintenance expenditure (Audit Commission 1988). Although we could probably buy our way out of winter chaos in a country like Britain, the cost would be extremely high. The question for the community at large is are we willing to do so?

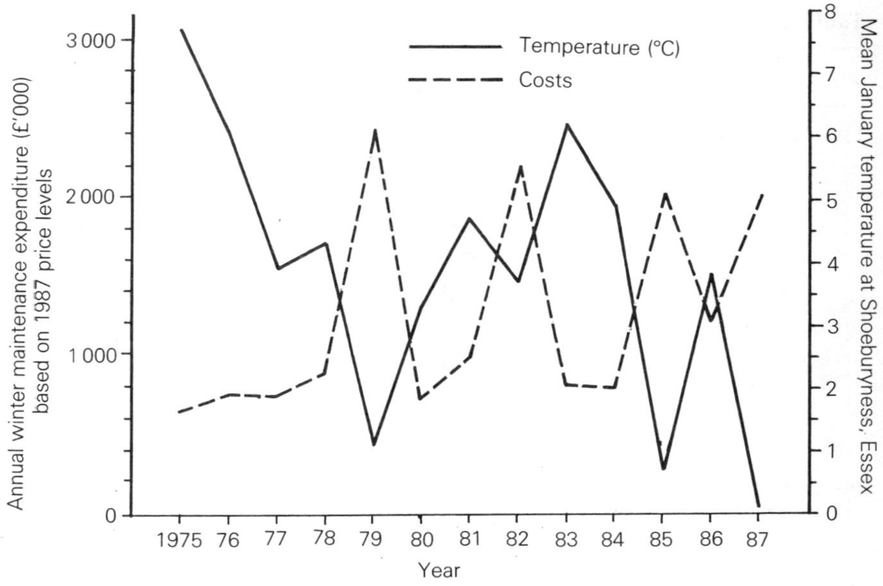

Figure 2.2 Annual winter maintenance expenditure in Essex for the period 1975–1987, and mean January temperature. Expenditure varies considerably, depending on winter severity.

Figure 2.3 Snow cleaning operations in city streets in progress.

2.5 REFERENCES

Audit commission (1988). *Improving Highway Maintenance*. London: HMSO.

Collier, C.G. and Chapuis, M. (eds) (1990). *Weather Radar Networking*. Seminar in COST Project 73, Dordrecht, Kluwer, for Commission of European Communities.

Gardner, D. (1987). Winter chaos – can we buy our way out of it? Introductory notes to conference on *Winter Chaos*, pp. 12–16. London: Institution of Civil Engineers.

Manson, J. (1987). Counting the cost of Britain's big freeze. *Highways*, 41–42.

Perry, A.H. (1981). *Environmental hazards in the British Isles*. London: George Allen & Unwin.

Perry, A.H. and Symons, L (1980). Economic and social disruption arising from the snowfall hazard in Scotland – the example of January 1978, *Scot. Geog. Mag.*, **96**, 20–25.

Perry, A.H., Symons, L. and Symons, A. (1986). Winter road sense. *Geographical Magazine*, **58**, 628–631.

Ponting. M. (1984). Weather prediction systems. *Highways & Transportation*, **31**, 24–32.

Symons, L. (1980). The economic and social disruption arising from the snowfall hazard in Scotland – the example of January 1978. *Scottish Geographical Magazine*, **96**, 20–25.

Chapter three

Thermal mapping and road-weather information systems for highway engineers

John E. Thornes

The 1980s have seen a technological revolution in the measurement and prediction of road surface conditions with the advent of such techniques as thermal mapping, road sensors and ice prediction.

The close co-operation between highway engineers and meteorologists that has developed in this time has been mainly concerned with the winter maintenance of roads: i.e. keeping roads as safe as possible by keeping them free of ice and snow. Not only does better weather information save maintenance costs, but also the introduction of sophisticated road-weather prediction systems into highway departments has provided a catalyst for rationalizing existing maintenance procedures. It is also now possible to provide road-weather information to other emergency services, utilities and of course the general public. Individual countries have developed their own systems to fulfil perceived road-weather requirements. This chapter attempts to provide an overview of recent developments in road-weather monitoring and communications primarily for winter maintenance activities, calling upon examples from different countries as appropriate. Table 3.1 reviews briefly the state of play in various European and other countries.

The general climate in Europe is such that most countries have snow and ice problems, even as far south as parts of Greece and Spain. There is a need for a winter index that can distinguish between winter maintenance requirements in different countries. The length of the winter season varies with latitude, altitude and distance from the coast, but it also varies from year to year in severity as well as in length. Figure 3.1 shows an example of a temporal winter index for

Table 3.1 *Brief summary of road-weather information systems in different countries (as of January 1991)*

Country	*Thermal Mapping* (km)	*Road-weather out stations* (no.)	*Computer weather network*
Austria	0	250 (?)	None
Belgium	0	0	None
Denmark	500	85	Yes
Finland	0	100	Yes
France	550	100	Yes
Germany	0	100	Yes
Holland	1800	75	Yes
Italy	0	25	None
Luxembourg	100	0	None
Norway	100	20	None
Spain	0	0	None
Sweden	12000	600	Yes
Switzerland	100	180	Yes
United Kingdom	32500	485	Yes
USA	500	200	Yes
Canada	400	10	Yes
Japan	0	30 (?)	None
Totals	48550	2260	

Note those marked with (?) are unconfirmed

Manchester Airport, which is discussed in detail elsewhere (Thornes 1988; Hulme 1982). Such an index compiled each winter across Europe and North America, could provide a useful comparison of potential and actual winter maintenance activities.

Zero degrees Celsius is an important threshold; as well as determining the possibility of frost or ice, the air temperature determines whether or not precipitation is likely to fall as snow. There are not yet sufficient road-surface temperature measurements to construct a map of minimum road surface temperatures, but it is the job of the weather forecaster accurately to predict the severity and timing of air and road-surface temperatures falling below zero. Ice is at its most slippery at 0 °C as shown in Figure 3.2 (Moore 1975), and hence marginal areas where the road temperature may just fall to zero for an hour or so may present a greater problem to the highway engineer than stretches of road that are well below zero.

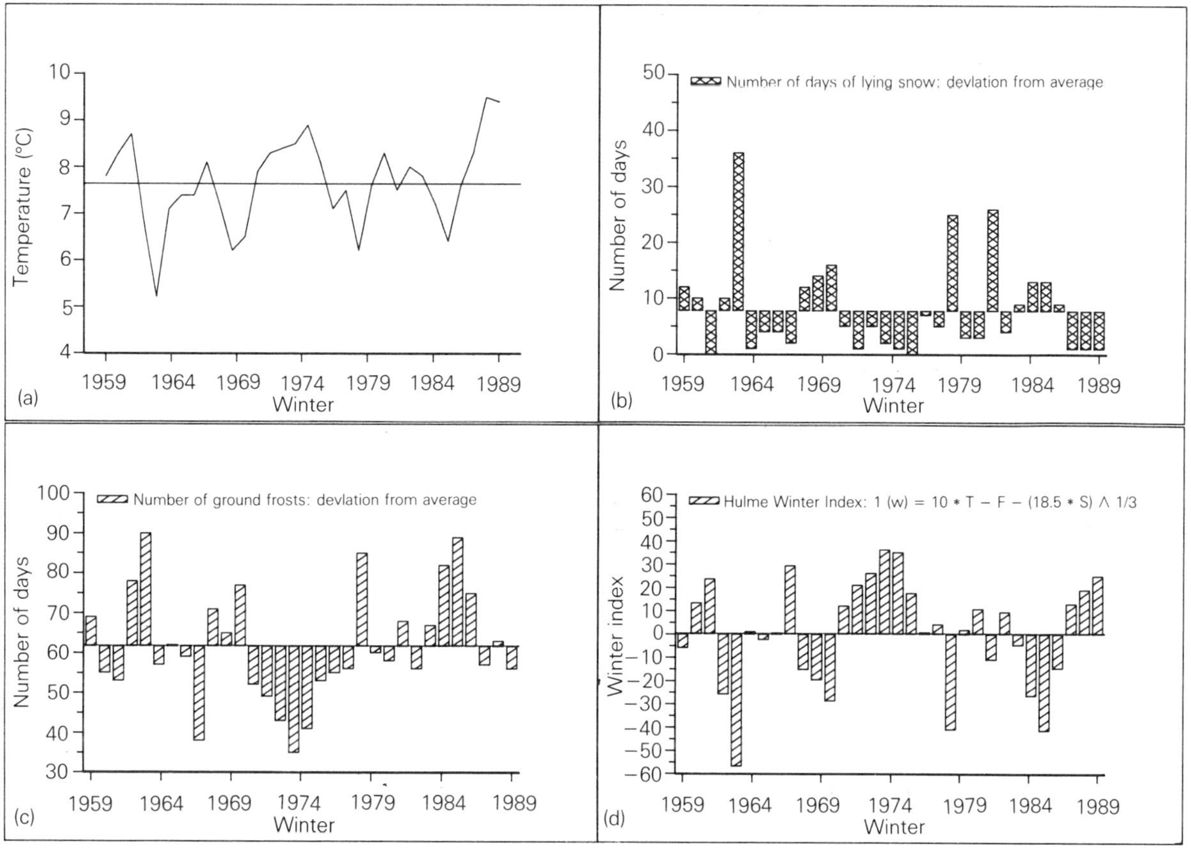

Figure 3.1 Temporal weather index for Manchester Airport (a) Mean maximum temperature November–March, Ringway; (b) Number of days of lying snow at 0900 GMT, Ringway; (c) Number of ground frosts, Ringway; and (d) Hulme winter index for Ringway.

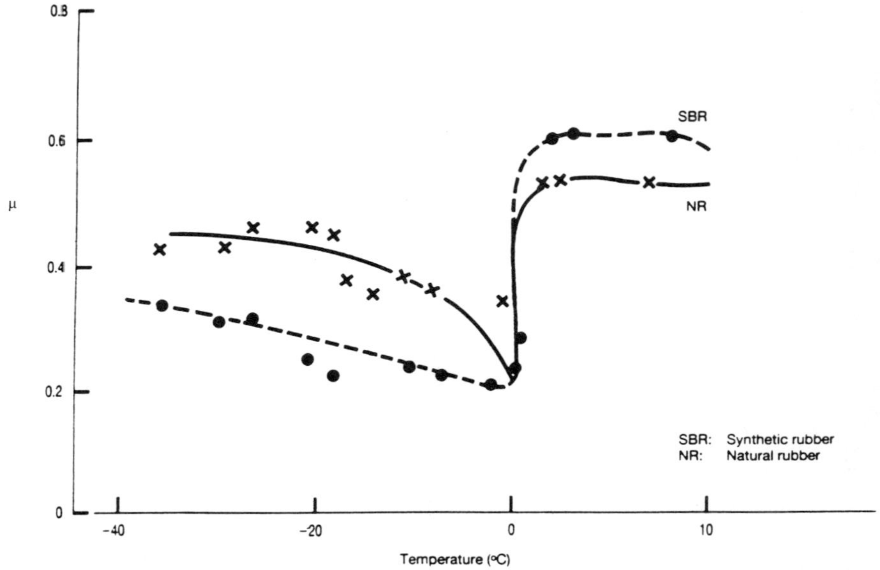

Figure 3.2 Skid resistance as a function of temperature (Moore 1975).

3.1 ROAD-WEATHER INFORMATION SYSTEMS

The main purpose of a road-weather information system is to reduce the cost of keeping roads free of ice and snow whilst at the same time making them easier and safer to travel on. This is achieved by linking together information about road surface temperature, wetness and residual chemicals with information about the synoptic weather conditions in order to give the highway engineer a prediction as to whether or not the roads will need to be treated in the next 24 hours or so. Applying salt before ice or snow occur requires considerably less salt: typically 10 g/m^2 of salt to prevent the formation of ice or the bonding of snow to the road, as opposed to approximately 40 g/m^2 to melt ice or snow. However, the salt must not be spread too soon or it may be carried off the road by traffic or precipitation. Normally salt should be spread within one to three hours of the likely formation of ice or the accumulation of snow.

There are four main components of an integrated road-weather system:

(a) spatial analysis of road microclimate, via thermal mapping;
(b) road and atmospheric sensors for real-time road-weather information;

(c) computer and communication networking;
(d) road-weather forecasts: snow and ice prediction.

All this information has to be presented to the highway engineer in a way that can be easily assimilated and acted upon. It is also sensible to keep a statistical record of the road-weather conditions and subsequent maintenance actions, for management appraisal of effectiveness.

3.2 ROAD MICROCLIMATE

Road surface conditions vary both spatially and temporally within a given road network. The technique of thermally mapping the spatial variations of road-surface temperature has been developed independently in Sweden and the United Kingdom. On calm clear nights the variations in road-surface temperature are due to altitude, topography, road construction and traffic. Wind, cloud and precipitation tend to reduce the amplitude of spatial variation in temperature. It is possible to construct a number of thermal maps of a road network that reflect the likely spatial variations in road-surface temperature on a given night. A meteorologist can choose the appropriate thermal map each night, and deduce its spatial applicability.

The measurement of road-surface temperatures is most easily achieved with an infra-red thermometer. The temperature is normally recorded in degrees Kelvin with a resolution of 0.1 K. Thermal mapping surveys are normally carried out when the average road surface temperature is below +5 °C.

3.2.1 Terminology

(a) *Thermal mapping*: The measurement of the spatial variation of road surface temperature along a road using an infra-red thermometer or camera.
(b) *Thermal fingerprint*: The graphical representation of temperature (y-axis) plotted against distance (x-axis) for a particular route on a given night. For an example see Figure 3.3.
(c) *Thermal map*: The representation on a road map (usually 1 : 50 000) of the average spatial variations of minimum road-surface temperature for different weather conditions.

3.2.2 Temporal variations of road surface temperature

It is necessary to carry out thermal mapping during the night. The diurnal rhythm of road-surface temperature is one in which the max-

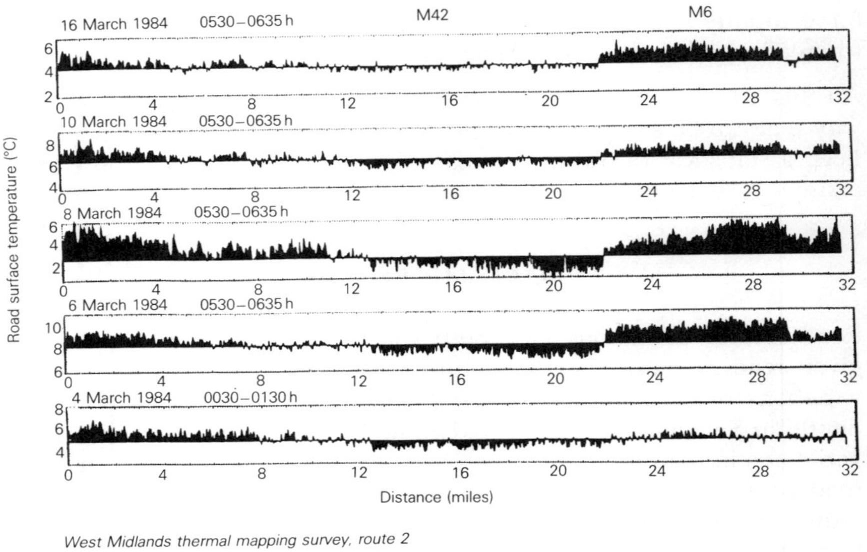

Figure 3.3 Sample thermal fingerprint: West Midlands thermal-mapping survey, route 2.

imum temperature normally occurs in the early afternoon and the minimum temperature occurs around dawn. Immediately after sunset, road-surface temperatures fall very rapidly but this decline levels off, so that during the latter part of the night they change very little. It is therefore possible to compare directly temperatures measured on one section of road with those recorded on another. This comparison is facilitated by the use of short routes (both in distance and time) in order to reduce the amount of surface temperature change between surveys.

The general diurnal temperature regime is affected by the prevailing weather conditions which influence the time of the maximum and minimum road surface temperatures and also the amplitude of the diurnal change.

Incoming solar radiation varies throughout the winter, in proportion with length of daylight and the height of the sun in the sky. Minimum solar input occurs on the shortest day (21 December), but the actual incident solar radiation at one place is also dependent on cloud cover and sky-view factor. Cloud cover reflects and absorbs solar radiation, thereby reducing the amount of direct solar radiation. However, clouds also re-emit, at longer wavelengths, absorbed solar radiation and reflect incident solar radiation to the surface. *Sky-view factor* is a term

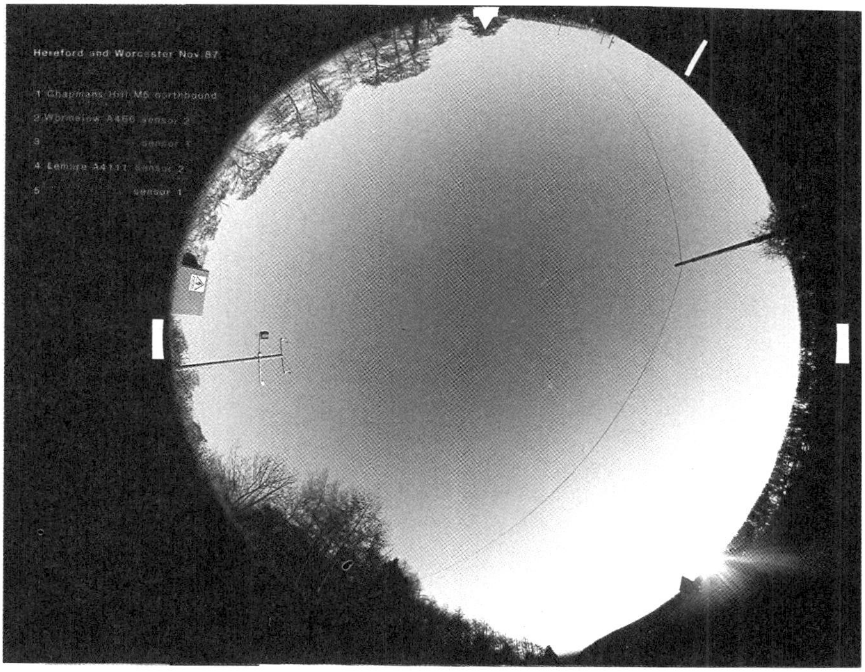

Figure 3.4 Fish-eye photograph of a site in Hereford and Worcester, showing a partially obstructed sky view, and the atmospheric sensors.

which has been used to relate the theoretical maximum incoming solar radiation to the actual figure. It varies from 0.0, when none of the sky is visible to the surface, to 1.0 when there are no obstructions. Generally, the sky-view factor depends upon tree and building cover, which reduce the incoming solar radiation by shading road surfaces. Thus, the solar radiation input not only varies seasonally but also daily depending on weather conditions, the amount of tree cover, topography and buildings. The sky-view factor also controls the loss of long-wave radiation from the road surface at night. Figure 3.4 is a fish-eye photograph of a site in Hereford and Worcester which shows a partially obstructed skyview and the atmospheric sensors.

The combined effect of buildings, topography and tree cover is generally systematic at a single point, although the variation of the height of the sun in the sky and the angle of incidence of the incoming solar radiation mean that the influence of the sky-view factor will change with season. The effect of tree cover on the sky-view factor will also be influenced by the amount of foliage left on the trees in winter.

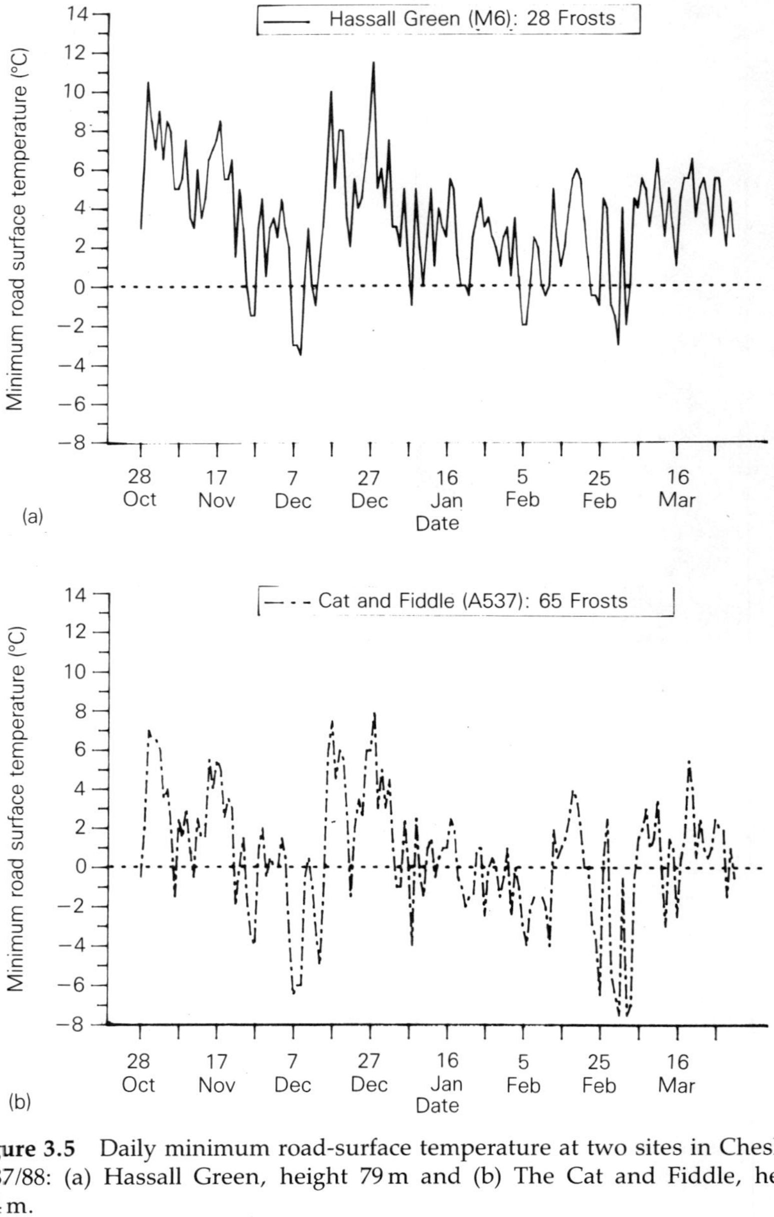

Figure 3.5 Daily minimum road-surface temperature at two sites in Cheshire, 1987/88: (a) Hassall Green, height 79 m and (b) The Cat and Fiddle, height 514 m.

The shedding by deciduous trees of their leaves will allow more long-wave radiation to escape at night, increasing the rate of cooling.

3.2.3 Systematic spatial variations

(a) Altitude

Normally, the higher the altitude the lower the road minimum temperature expected. This is the result of the decrease in air temperature with height which occurs in a normal, unstable atmosphere. The environmental lapse rate (the fall of air temperature with height above sea level) is usually about 6.5 °C/1000 m. Road-surface temperature could be expected to decline with altitude at a similar rate. Figure 3.5 shows a comparison of daily minimum road surface temperature at two sites in Cheshire: Hassall Green on the M6 at 79 m and the Cat and Fiddle at 514 m. The mean temperature difference is 3.4 °C for a height difference of 435 m giving a lapse rate of 7.8 °C/1000 m.

However, frost hollows in valleys at low altitude can cause the lowest temperatures to be recorded in valley bottoms, especially on clear and calm nights due to either the formation of inversions or the pooling of cold air. Inversions occur under clear calm conditions. The ground cools the air immediately above it so that air temperature increases with height. Above the inversion layer the normal decrease of air temperature with height resumes.

Cold air, being denser than warm air, will tend to fall under the influence of gravity. This is called *katabatic drainage*. A frost hollow will occur where a hill slope is sufficiently steep for drainage to take place, resulting in lower road temperatures.

The size of the cold air pool is related to the length and steepness of the slope. A cold air pool at the bottom of a long, shallow slope will be greater in extent than one at the bottom of a short, steep slope because of the greater volume of cold air on the longer slope. However, the cold air at the bottom of the short, steep slope may experience colder temperatures due to the greater difference in height and therefore in temperatures between top and bottom of the slope.

In some circumstances, where there is sufficient relative relief, the normal lapse rate will give cold temperatures at the higher altitude and cold air drainage will also give low temperatures at lower altitudes. In these situations, the warmest temperatures are obtained at middle altitudes between the cold hilltops and the cold valley bottoms. The phenomenon of warmer temperatures at middle altitudes is referred to as the *thermal belt*.

Figure 3.6 shows the variation of minimum air temperature with altitude for twelve sensor sites in Hereford and Worcester under the

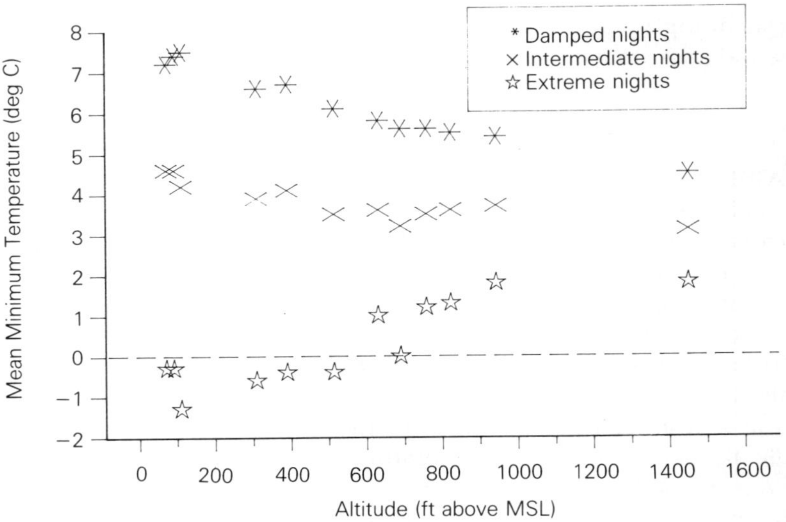

Figure 3.6 Variation of mean minimum air temperature with altitude for twelve sensor sites in Hereford and Worcester, 1988/89. The three weather types are discussed in section 3.1.1(d).

three weather types as discussed below in section 3.1.1 (d). The coldest air temperatures occur in valley bottoms in calm clear conditions.

(b) Topography

Topography restricts radiational cooling of a road surface, limiting the amount of long-wave radiation that can escape from the road by controlling the sky-view factor. At night, a road surface cools by radiation loss.

Radiation loss to the environment is reduced by buildings, trees, cloud cover, traffic and cuttings, all of which reflect, absorb and re-emit back to the surface, thereby restricting radiation loss from the road surface and maintaining temperatures. Hence roads in cuttings, under bridges, or lined by trees and/or buildings will stay warmer at night than more exposed roads.

Conversely, it must be remembered that sheltered roads may warm more slowly than more exposed roads since the early morning solar radiation cannot reach the road surface. This can be important: consider a night of moderate overnight frost during which hoar frost has sublimed onto the road surface and surrounding fields. After sunrise, these hoar-frost deposits are melted or sublimated by the incident solar radiation on exposed road sections. In shaded sections where solar radiation is unable to penetrate, road-surface temperatures may remain low and early morning traffic can then compact the hoar frost into ice. Hence areas with a low sky-view factor can be more hazardous than exposed sections if the road-surface temperature falls below zero.

(c) Road Construction

Road construction is important because heat is stored in the road structure and released differentially according to its thermal properties. Depth of construction is important too: usually the greater the depth of construction the warmer the road. As a result, motorways are normally warmer than other roads, and concrete roads are warmer than blacktop roads. Also, the seasonal variation in incident solar radiation must be considered. In late Autumn and Spring when frost formation is still a hazard, sufficient radiation may be stored in the road from the daytime input to offset the night time cooling.

Where a road crosses a bridge it is likely to be colder due to its shallower construction and, as a result, smaller 'thermal memory'. This term is used to describe the length of time which a road structure-or any structure-retains the stored heat which it gains from daytime solar radiation. The thermal memory depends upon the depth of construction, the construction materials used and the amount of incident solar radiation received. Certain bridges, particularly those over water, may appear warmer as a result of radiation to the underside of the bridge from a relatively warm water surface. In urban areas, elevated viaduct sections, although limited in depth of construction, can remain warm due to the effect of the urban heat island (see below) and traffic. Steel bridges normally cool quicker than the adjacent road thanks to their high thermal conductivity and poor heat retention. They present particular problems as they can produce a short icy section on an otherwise safe road.

The construction of major roads such as motorways and trunk roads tends to reduce the effects of minor topographical features. For example, embankments reduce the effect of cold-air pooling by raising the road above the base of frost hollows and valleys. Cuttings reduce altitudinal variations and also reduce the sky-view factor.

(d) Urban Heat Island

The urban heat island effect is the phenomenon observed in towns and cities where the built-up area can be several degrees warmer than the suburbs or surrounding rural area. The actual magnitude of the heat island at any single location in the city will depend upon the season and land use at that location. Urban heat island intensity is a function of city size, population density and urban morphology. The temperature difference is due to the industrial and domestic heat sources, the low sky-view factor and heat retention by the fabric of the city.

The urban heat island effect is strongest in the autumn and spring, and is at its weakest in the midst of winter. The effect is most noticeable when wind speeds are low, reducing mixing of the air over the city. Within the city the heat island effect means that the effects of topography, weather and traffic are usually less influential on road surface temperatures than on non-urban roads.

(e) Traffic

Traffic tends to keep a road warm at night by acting as a shadow factor restricting the loss of radiation. Also traffic stirs the air above the road surface, mixing in warmer air from above on cold nights. In addition radiation from the engine and exhaust plus the frictional heating from tyres mean that minimum temperatures can be up to two degrees warmer than for an untrafficked road. On multi-laned roads vehicles tend to concentrate on the 'slow' lane which means that the 'fast' lane and slip roads may well be cooler at night.

3.2.4 Observed thermal fingerprints in differing weather conditions

(a) Weather

The weather conditions also influence the spatial variations of road-surface temperature. The thermal fingerprint of minimum road surface temperature is most clearly developed on clear, calm nights, typical of anticyclonic (high-pressure) weather systems, when temperature inversions are most marked. Rain and wind tend to dampen the fingerprints as discussed below.

(b) Extreme thermal fingerprints

Road surfaces on anticyclonic nights are often dry but there is the danger of frost formation: frost which may be compacted into ice by the pressure of traffic. Hoar frost will be formed on a road surface when a parcel of air containing water vapour comes into contact with a road surface whose temperature is below 0 °C and below the frost point of the water vapour. Calm, clear nights produce the maximum variation in road-surface temperature. Cold-air pooling in frost hollows can lead to low road temperatures in valleys.

(c) Damped thermal fingerprints

If roads are wet, then the amplitude of the thermal fingerprint is reduced-literally and metaphorically damped. Overcast, rainy and windy nights, which are typically associated with cyclonic weather, give smaller variations in road-surface temperature. High wind speeds promote mixing of air layers, preventing the formation of inversions. Cloud cover reflects, absorbs and re-emits long-wave terrestrial radiation emitted by the road surface, reducing the amount of radiational cooling at the road surface, and hence keeping surface temperatures higher than under clear sky conditions.

Depressions are not always related to ice formation in winter. However, one important scenario occurs after the passage of a cold front, producing rain, followed by clearing skies. All the wet roads in an area will start off at a similar temperature. The rate of evaporation and cooling will depend upon the topography, but the stretches of road that fall to 0 °C most quickly will be the same stretches of road that are cold on clear nights.

(d) Intermediate thermal fingerprints

Roads at higher altitudes will be colder than roads at low altitudes on clear but windy nights when cold-air drainage or inversions do not occur. Due to the environmental lapse rate we normally expect upland roads to be colder, but upland areas are more susceptible to the prolonged effects of cloud and rain than lowland areas, so the coldest site on average in a region may be at low elevation. On windy nights, however, the coldest sites are usually those at highest elevation.

Most nights give rise to intermediate thermal fingerprints which are somewhere between the extreme and damped thermal fingerprints depending upon the amount of wind and cloud.

(e) Other fingerprints

Obviously, other factors have to be taken into account on nights when the weather conditions change. Patchy fog results in local variations in the thermal map, and freezing fog will lower road surface temperatures. Similarly, broken, stationary cloud will limit profile development and cause local variations.

(f) Summary of fingerprint development

The fingerprint is most developed on clear nights with little wind; under other weather situations it is less pronounced. On any given night the maintenance engineer will wish to have sufficient suitably sited sensors, or weather information, to give him an impression of the sort of thermal fingerprint that has developed. However, the size of temperature variations along the stretch of road – i.e. the amplitude and shape of the thermal fingerprint – depends upon the geographical make-up of the road network.

(g) Forecast thermal maps

The thermal fingerprints can be related to a map of the road network, and on individual nights a map of expected minimum road surface temperature can be produced. Such a forecast thermal map is issued by combining the forecast minimum road-surface temperature from a weather office with the appropriate thermal map.

Owing to the geography of a region, zonal climates can be identified such that a different thermal map can be chosen for different zones on a given night. In the county of Cheshire in England, three regional climates have been identified, each with three thermal map types. For example, on a given night there are 27 ($3 \times 3 \times 3$) possible combinations of thermal map for the county. The larger the region, and the more varied the geography, the greater the variations. Variable climatic domains can be presented according to the weather conditions.

This has considerable significance for the maintenance engineer. The likely spatial severity of weather conditions can be related to road-surface conditions and presented to the engineer in advance of their occurrence, enabling resources to be mobilized more effectively.

Recent developments include the use of aerial (Beaumont *et al.* 1987) and satellite survey techniques (McClatchy *et al.* 1987), but both methods are expensive and are restricted to cloud-free operation.

In both Sweden (Olafsson 1985) and the United Kingdom (Thornes 1985) thermal maps of road surface temperature are produced using vehicle-mounted infra-red thermometers. Bogren (1990) and Gustavsson (1990) have quantified many of the relationships discussed above. Figure 3.7a shows the counties that have carried out thermal-mapping surveys in Great Britain and Figure 3.7b shows the countries that have carried out thermal mapping up to the end of 1990.

(h) The use of thermal maps

Thermal mapping is of most use in climates where the minimum road temperatures in winter are close to 0 °C. Forecast thermal maps can help to identify which parts of a road network need winter maintenance treatment first. In an ideal world, thermal maps can be used to redesign 'gritting' routes so that only 'cold' routes are treated on marginal nights, or so that cold spots, such as bridges, are treated first.

Thermal mapping is also used to locate permanent road sensors, and to reduce the number of road sensors required in a region. In the future it may be possible for gritting lorries to put more de-icing chemical down on cold stretches of a route, the chemical spread rate being controlled by the thermal map stored in an on-board computer.

(a)

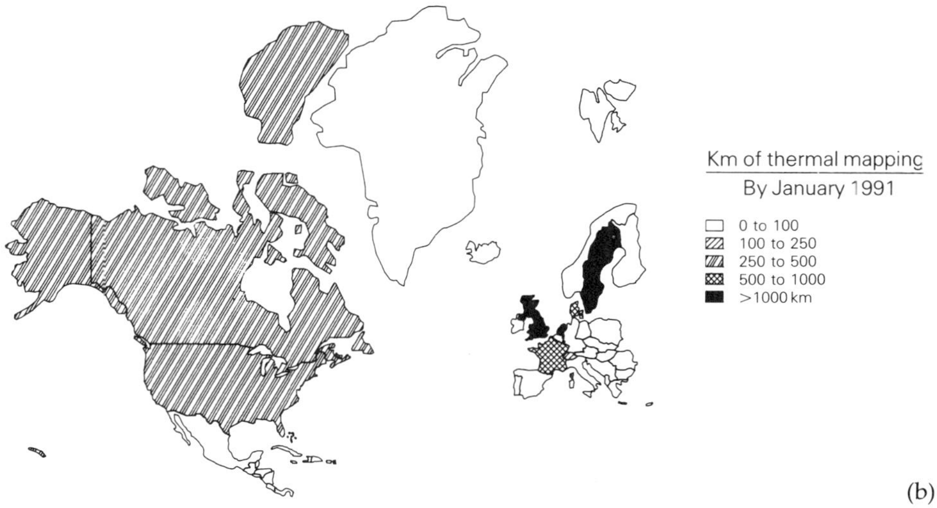

(b)

Figure 3.7 (a) Kilometres of thermal mapping in the UK and (b) Kilometres of thermal mapping by country.

Table 3.2 *Possible errors in vehicle-based thermal mapping*

1. Emissivity of road surface
   The emissivity of the instrument can usually be varied, but an error of 1% = 0.5 deg.C (e.g. 0.95 instead of 0.96)
2. Temperature of instrument
   Infra-red thermometers are normally calibrated to operate within a certain instrument temperature range, this range may be exceeded
3. Detector sensitivity
   Instrument detectors will all respond slightly differently to give different signal/noise ratios
4. Millivolt to temperature conversion
   Errors in method e.g. regression
5. Dirty lens/condensation
   The lens of the instrument may become dirty due to splash and/or condensation may take place on the lens
6. Atmospheric absorption
   Not all radiation leaving the road surface will arrive at the instrument
7. Vehicle radiation
   Radiation from the vehicle exhaust and/or engine may enter instrument
8. Lens waveband
   The lens is designed to transmit only a part of the infra-red spectrum
9. Tyre pressure
   The distance covered between observations may vary between sampling runs due to differences in tyre circumference
10. Lane changes
    The outside lane will be coldest due to less traffic, but most surveys are carried out on inside lane. Vehicles may have to move into different lanes on different surveys.
11. Warm-up of instrument
    Some instruments require to be switched on up to an hour before use

Note: Systematic errors can be eliminated by careful preparation (e.g. 2, 4, 5, 7, 9) but the others may be random between surveys and different instruments.

(i) Limitations of thermal mapping

Thermal mapping only produces relative minimum temperature differences between stretches of road. It can be used to measure actual road surface temperature accurately, only to within ±2 °C. This is because the emissivity of roads is not known accurately. If the emissivities of the road surfaces in an area are different then care must be taken in interpreting the results. Also, a dry road will have a different emissivity from a wet or snow-covered road. A difference in emissivity of 1% can give a temperature difference of 0.5 °C. Again, care must be taken to maintain the infra-red camera within a constant temperature range during observations, and the infra-red cameras must be regularly calibrated. Table 3.2 lists the likely sources of error for infra-red measurements. There is a UK Department of Transport specification for thermal mapping which sets out minimum standards for contractors (Department of Transport 1988b).

3.3 ROAD SENSORS

3.3.1 Active versus passive versus non-contact

There are three main types of road sensor: active, passive and non-contact. *Active sensors* attempt to predict whether ice is likely to form on a road surface by cooling an area on the surface of the sensor by approximately 2 °C below ambient. If ice or frost is detected then a warning is given. Other developments of active sensors include a heated area to detect dry snow or ice, and a variably cooled area to detect the freezing temperature of any surface moisture.

Passive sensors just sit in the road surface measuring road surface conditions without adding or taking away energy from the system. Both types of sensor also attempt to measure surface moisture and residual de-icing salt.

Non-contact sensors include the use of microwaves and infrared, using sensors mounted on gantries or poles at the side of the road. More research is required to make them operationally cost-effective.

Road sensors are now quite reliable for measuring road-surface temperature, (Table 3.3 gives a list of likely errors), but there is still much room for improvement in the measurement of surface moisture and residual de-icing salt. If the road surface is wet due to precipitation then the measurements are more reliable, but if the road is dry then measurements of residual de-icing salt are impossible. Also, problems arise because many de-icing agents are hygroscopic, causing the road to appear wet to the sensor which gives the engineer a false impression that more salt is required to treat the wet surface.

Most of these problems can be overcome by relating the sensor readings to the weather conditions, and by studying the displayed chemical factor readings over time.

3.3.2 Location and number

The number of sensors required adequately to cover a road network, and the correct location of the sensors, can be assessed by the use of thermal mapping. Normally a power supply and phone line are required at each sensor outstation and often it is expensive to supply these utilities to remote sites. Thermal mapping enables more con-

Table 3.3 *Possible errors in the measurement of road surface temperature by surface sensors*

1. Thermal properties of sensor materials
   Thermal conductivity, thermal capacity, air voids
2. Thermal properties of backfill materials
   Thermal conductivity, thermal capacity, air voids
3. Depth of temperature sensor below surface of road (mm)
4. Analogue to digital conversion
   8, 12 or 16 Bit conversion, sampling technique: single shot or multi-sample, background signal noise
5. Calibration of temperature sensor
   Operating range
6. Albedo of sensor if different to that of road
   Colour coding not always adequate – leads to diurnal variations in accuracy
7. Emissivity of sensor if different to that of road
   1% difference = 0.5 deg.C (e.g. 0.95 instead of 0.96)
8. Seasonal systematic errors
   Thermal shorting more important when large temperature gradient in ground (autumn and spring) – diurnal and seasonal effects
9. Sensor wet or dry
   Sensor calibration normally carried out when sensor wet – better thermal contact with road when wet
10. Sensor polling time
    If there are many sensors in a system it may take up to half an hour before last sensor is polled, and yet data is stored as if it was recorded on the hour

Note: Systematic errors can be eliminated or minimised by careful design of the sensors and associated communication systems (e.g. 1, 2, 3, 4, 5, 6, 7, 10), but the other errors may be random and are more difficult to deal with.

venient sites to be chosen, as the road conditions at the remote sites can be inferred from the thermal map. A network of sensor sites is required across a region: normally one sensor site per 250 km^2 depending upon road density and regional climatic variations. In the near future battery-powered sensor systems with satellite communications will improve the reliability of road-weather out-stations.

It is sensible to place sensors in a variety of microclimatic locations so as to avoid over-pessimistic or optimistic results. For instance, placing sensors only at cold spots gives a very pessimistic view of the road conditions across a region.

3.3.3 Sensor accuracy maintenance and calibration

Normally at a typical outstation several road surface sensors are linked to a single set of atmospheric sensors which measure wind speed and direction, air temperature and humidity, precipitation and perhaps radiation. The atmospheric sensors are often considered optional by the engineers, but are very useful to the meteorologist.

The temperature measurements are normally expected to have an accuracy of 0.5 °C, with a resolution of 0.1 °C. The wind measurements are expected to be within ±5% accuracy, and the relative humidity within ±2%. It is very difficult to check the accuracy of the sensors without regular calibration and maintenance visits. Humidity is particularly difficult to measure accurately (especially automatically), and monthly calibration is ideally required. Preventive maintenance is ideal, and a compromise is to calibrate all the sensors every two months in winter with a minimum of once at the start of the winter and a second in January or February. The Department of Transport protocol MCE2020G (Department of Transport 1988a) gives minimum standards for sensor performance. Road sensors should ideally be calibrated just before dawn on a cloudy night with a wet road surface to reduce measurement errors.

Automatic self-calibration checks are now possible for temperature measurements which may speed up calibration visits, and it is hoped that eventually all the sensors will be self-calibrating.

An international standard for road/atmospheric sensor accuracy may be forthcoming from the COST 309 programme, so that manufacturers of equipment are competing equally. A more difficult matter is the policing of such standards: certificates of compliance provided by manufacturers provide one solution.

Figure 3.8a shows the highway authorities that have installed road weather outstations in the UK, and Figure 3.8b shows those countries that had installed outstations up to the end of 1990.

3.4 COMPUTER AND COMMUNICATION NETWORKS

Information from sensors and weather offices has to be combined and distributed to the engineer, other emergency services and the general public if required. Several countries have experimented with some form of teletext service to distribute the information as widely as possible. As yet teletext graphics and communications are slow but no doubt improvements will prove this to be a popular solution. Teletext systems normally allow information to be displayed only, and not manipulated, unless the information is downloaded onto a computer. This has meant the development of dedicated computer systems and networks for road weather, leading to a proliferation of terminals on an engineer's desk, unless existing computer equipment has been

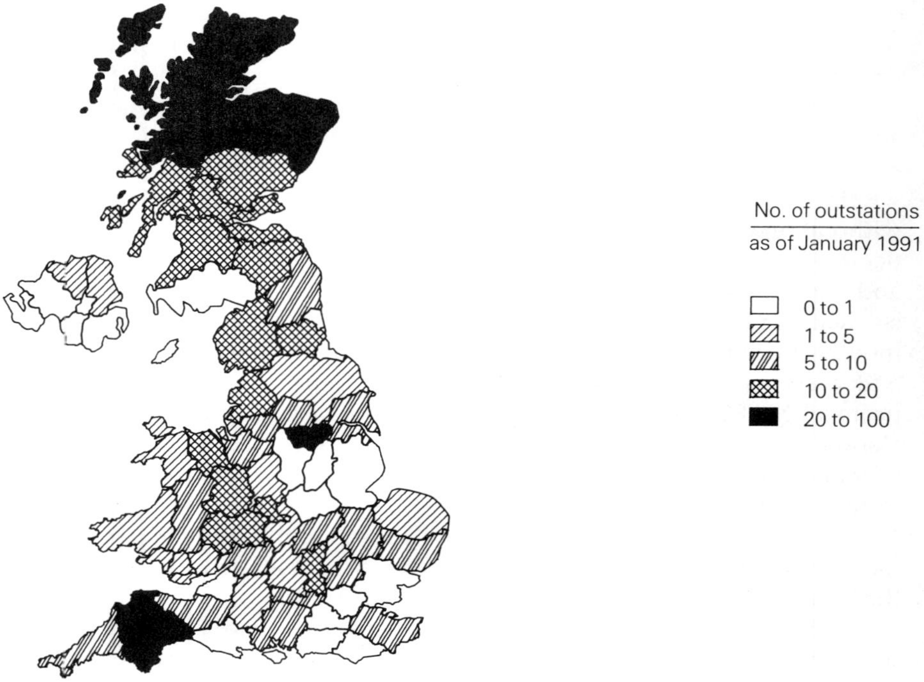

(a)

(b)

Figure 3.8 (a) Number of outstations in the UK by county/region and (b) number of outstations by country.

utilized. Micro-computers are obviously popular in that they can be used for a variety of purposes: terminals to minicomputers or mainframes, stand-alone word processors/report generators, and terminals to a variety of applications such as road-weather, and other highway maintenance activities. These 'workstations' as they have come to be known, can also access teletext, act as docfax/telex machines etc. if required. The problem is one of universal compatability, although IBM compatability has gone a long way towards solving this problem. To use a road-weather system requires the engineer to become familiar with computers but this is less of a problem than it was.

Most road-weather systems require the highway authorities to have their own central processor unit, or central station, to collect the sensor data at regular intervals. This is normally accomplished via a modem link using conventional telephone networks. In certain circumstances it may be possible to use a radio link, especially for small line-of-site systems. In the future satellite communication will be common as it becomes more cost-effective.

The central processor unit has to communicate the sensor information to the weather office and local workstations. A clearly defined communication protocol has been developed in the United Kingdom to enable weather offices to communicate with any commercially available sensor system. This has been in successful operation since 1986. More recent developments in the United Kingdom have eliminated the need for all authorities to have their own central processor units. At present in the United Kingdom each highway authority is responsible for its own commitment to ice prediction systems, but there are already signs of regional systems developing, as in Wales (Perry and Symons 1986). Also several computer bureaux have been installed in both weather centres and manufacturer's offices to facilitate communication between highway authorities and also to keep costs down.

3.5 ICE-PREDICTION MODELS

The Open Road service provided by the UK Meteorological Office is distributed to more than fifty counties/regions by fourteen weather centres. An ice-prediction computer model is run for forecast sites in defined climatic zones within each county/region. These forecasts are monitored to assess their accuracy, and to provide feedback to the forecaster as to how well the road surface temperature and wetness are modelled. Figure 3.9 shows the forecast accuracy for part of the winter of 1987/88 for the Ray Hall outstation in the West Midlands. Figure 3.10 shows the mean forecast accuracy in frost prediction (whether or not

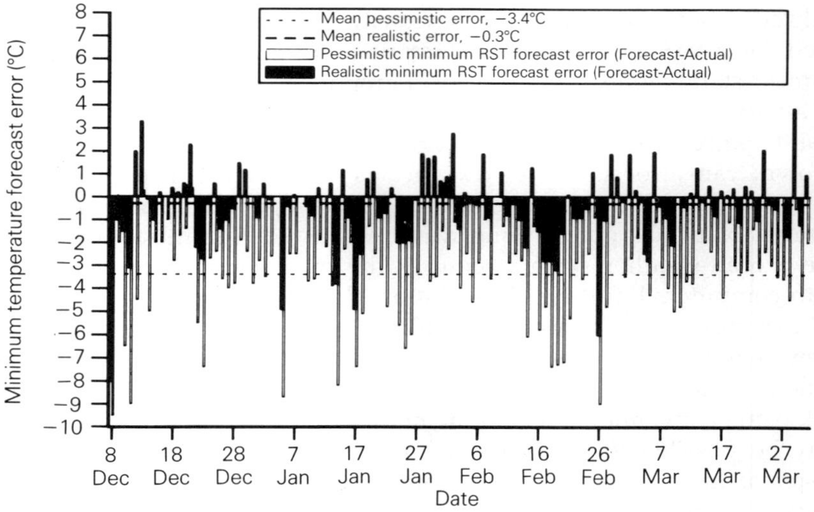

Figure 3.9 Forecast accuracy for the Ray Hall outstation in the West Midlands, winter 1987/88.

Table 3.4 *Forecast input sheet for road surface temperature model*

Thermal Mapping International Ltd	NATIONAL ICE PREDICTION MODEL
Site: Coleshill M42	Date: 10DEC88 REALISTIC
Road surface temperature: 6.2	
Road depth temperature: 3.7	

Time (GMT)	1200	1500	1800	2100	2400	0300	0600	0900	1200
Air Temp	5.7	7.0	4.5	2.0	0.5	−0.5	−1.0	1.0	4.0
Dew Point	1.3	4.0	2.0	−0.5	−1.0	−1.5	−2.0	−1.0	0.5
Avg Wind Speed	10	5	3	2	2	2	3	5	
Avg Cloud Amnt	4	3	2	2	2	2	2	2	
Avg Cloud Type	1	1	1	3	3	3	3	1	
Avg Rainfall	0	0	0	0	0	0	0	0	

the road-surface temperature fell below zero) for eight forecast sites in three counties, issued by Birmingham Weather Centre during the winter of 1988/89. Two forecast curves are issued, one realistic and one pessimistic. The realistic forecast is produced from the forecasters' best estimate of the likely weather conditions which are fed into the computer model as shown in Table 3.4. The pessimistic forecast repre-

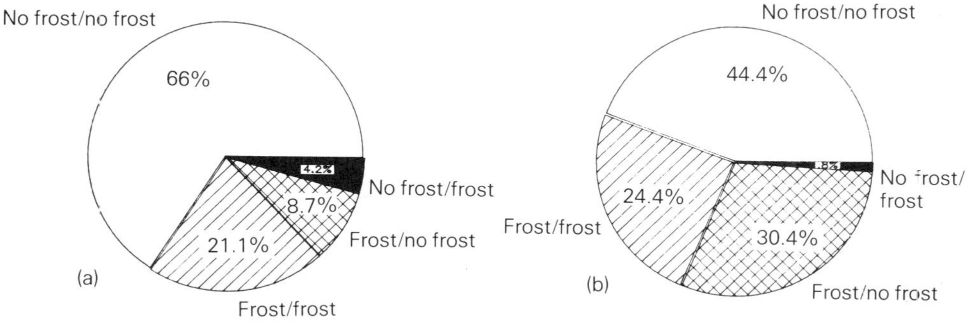

Figure 3.10 Mean forecast accuracy in frost prediction, Birmingham Weather Centre. Data compiled for eight forecast sites, in three counties, for winter 1988/89. Segment names: forecast condition/actual condition. (a) Realistic forecast. Percentage correct forecasts: 87.1% and (b) Pessimistic forcast. Percentage correct forecasts: 68.8%.

sents the worst scenario that is likely to happen on a given night. Figure 3.9 shows that the mean error in the minimum road-surface temperature forecast (for 113 nights) for the realistic curve was −0.3 °C, and for the pessimistic curve was −3.4 °C. Figure 3.10 shows that for the 151 nights considered, the accuracy of the realistic forecast in determining whether or not the road surface temperature would fall below zero was 87.1%; for the pessimistic forecast it was 68.8%. There are two types of error in the forecast that are highlighted in Figure 3.10. The shaded area represents those nights (8.7%) when frost was forecast but no frost occurred, and the solid area (4.2%) those nights when no frost was forecast but frost actually occurred. A frost in this context just represents road-surface temperatures falling below zero. The potential consequences of these two types of error are very different:

Type 1 error: No frost forecast/frost occurs: potential accidents
Type 2 error: Frost forecast/no frost occurs: potential wasted salt

Obviously one wants to reduce both types of error to a minimum but the Type 1 error is the more serious. Approximately 33% of the errors in the realistic forecast are Type 1 errors (4.2% out of 12.9%). If both curves were used when most appropriate the potential accuracy was 91.5%, with a type 1 error on just one night. (Thornes and Fairmainer, 1989; Thornes, 1989).

Further research (Thornes and Shao, 1991) is leading to an improvement in the ice prediction models, not just for road-surface temperature modelling, but also for improved prediction of the occurrence

Table 3.5 *Comparison of Retrospective (RSP) and 'Real-time' (RTP) Runs (Chapman's Hill, winter 1988/89, 65 days)*

	Icebreak		Thornes		Met. Office	
	RSP	RTP	RSP	RTP	RSP	RTP
No. of hours	1480*	1480	1480	1480	1480	1480
Overall:						
Bias (deg C)	−0.33	−0.10	−0.61	−0.17	−1.15	−0.42
SD (deg C)	0.75	1.14	0.97	1.12	0.81	1.02
RMS (deg C)	0.93	1.43	1.31	1.42	1.45	1.28
Maximum:						
Bias (deg C)	−0.56	−0.16	−1.10	−0.73	−1.13	−0.78
SD (deg C)	1.58	1.63	1.96	1.87	1.60	1.52
Minimum:						
Bias (deg C)	−0.17	−0.38	−0.19	−0.11	−0.94	−0.26
SD (deg C)	0.69	1.64	1.10	1.48	0.85	1.32
Time of Freezing:						
Freezing (nights)	17	17	15	15	17	16
Start (minutes)	37	64	−28	52	46	−49
Duration (minutes)	52	56	−12	68	103	0

* 1480 hours out of 1560 possible (95% availability)

and timing of 'wet frosts'. Two models have been used in the Open Road programme, *the Thornes model* developed at the University of Birmingham (Thornes, 1985; Parmenter and Thornes, 1986), and the Meteorological Office model (Rayer, 1987). Both models have been the subject of much recent research, and Table 3.5 shows results for 65 nights at Chapmans Hill on the M5 compared to a new model developed by Thornes and Shao (1991) called Icebreaker. Note that these results are for both retrospective modelling (RSP) using actual figures for air temperature, cloud, wind, humidity and precipitation observed at the University of Birmingham and for real time forecasting (RTP) using open road input data from Birmingham Weather Centre.

3.6 COST/BENEFIT OF ROAD-WEATHER SYSTEMS

It is always difficult to carry out an objective analysis of the benefits of a road-weather system. The costs of installation and maintenance are clearly easier to arrive at. The benefits of an effective winter-maintenance service, such as reduced travel time, reduced accidents, and reduced environmental damage, are difficult to quantify in the

Table 3.6 *Salt usage and winter index 1982/83 to 1989/90 Cheshire County Council*

	Winter index (Manchester)	*Salt usage (metric tonnes 000's)*
1982/83	10	17
1983/84	−5	19
1984/85	−27	24
1985/86	−41	27
1986/87	−15	22
1987/88	13	13
1988/89	19	11
1989/90	25	11

short term. Nevertheless, the few studies that have been carried out, for instance in Finland (Ministry of Communication, Finland 1982) and in the United Kingdom (Thornes 1989), suggest that considerable benefits are immediately apparent following the installation of a road-weather system.

3.6.1 Winter index versus salt usage in two UK counties

(a) Cheshire County Council

The temporal winter index derived in Figure 3.1 has been compared with salt-usage figures for eight winters in Cheshire County Council. Before these figures are compared it must be remembered that the winter index is for just one site which cannot be considered to be representative of the whole county. The salt data are only available as a county total, and obviously these totals are only best estimates of usage, correct to perhaps ±1000 tonnes as shown in Table 3.6.

Figure 3.11 shows that there is a very strong inverse linear relationship although the sample size is too small to make any firm conclusions. Nevertheless it is interesting to note that Cheshire installed their ice-prediction system during the 1986/87 winter, and it appears that although there was no obvious reduction in salt usage in the first winter, there was a real reduction in the last three winters. Predicting the 1987/88 to 1989/90 salt usage from the previous five years' data using regression analysis produces predictions of 16 000 and 14 900 and 13 700 tonnes respectively. The actual use of 13 050 in 1987/88 represents almost a 20% reduction, the 10 960 in 1988/89 represents approximately a 26% reduction and the 10 903 (in 89/90) represents a 20% reduction. The strength of the inverse relationship is remarkable,

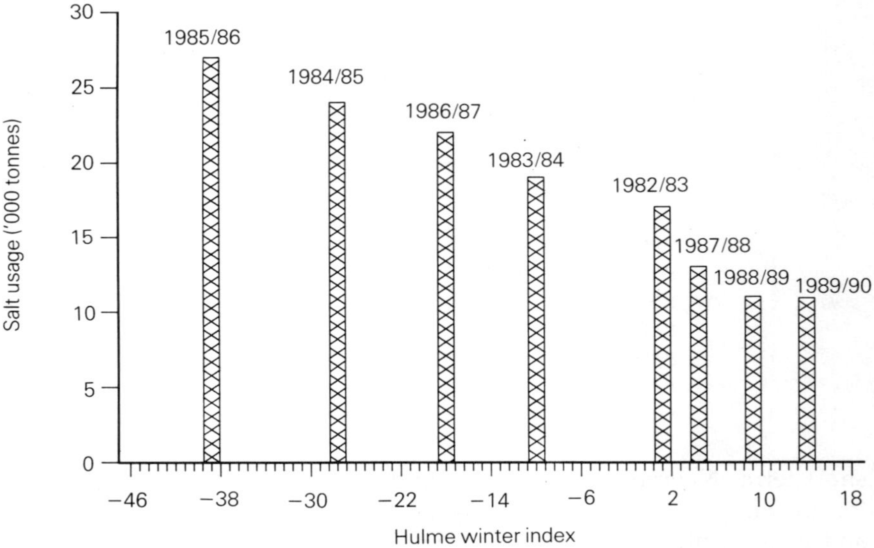

Figure 3.11 Salt usage versus Hulme winter index, Cheshire County Council.

but it is planned to monitor salt usage carefully in Cheshire for at least another two winters before firm conclusions can be drawn.

(b) Hereford and Worcester County Council

Ponting (1984, 1989) has published a detailed examination of salt usage in Hereford and Worcester County Council. He showed that during the winter of 1983/84 weather forecasts were only on average 57% accurate in the prediction of wet frosts, and that with hindsight a wastage of approximately 30% in salt usage was apparent. During the spring of 1987 Hereford and Worcester installed an ice-prediction system and an identical study to the earlier one was carried out during the winter of 1987/88. This represents a fascinating and unique 'before' and 'after' ice-prediction study in two winters that were not too dissimilar. Initial findings show that the accuracy of wet frost prediction went up to 91%, and that salt wastage was reduced to approximately 15%. Over the last ten years Hereford and Worcester have used an average of 18 620 tonnes of salt; a reduction in wastage of 15% would save on average 2800 tonnes a year. At a cost for salt of approximately £20/tonne in the United Kingdom this represents an average saving of £56 000. The total savings, including reduced labour costs and wear and tear on equipment, make the average annual savings much higher.

(c) United States – Strategic Highway Research Program
The Strategic Highway Research Program is a $150 million, 5 year project instigated in 1987. Project H207 entitled 'Winter Storm Monitoring and Communications' started in November 1988 and is investigating amongst other topics the cost/benefit of thermal mapping and road-weather information systems. Figure 3.12 shows the results of a questionnaire relating to the current cost of winter maintenance in the United States which totals nearly $2 billion each winter when cities, countries and states are all taken into account (Boselly *et al.*, 1990). The full results are to be published in 1992.

3.7 CONCLUSION

During the 1980s road-weather information systems have come of age in most of Europe in a range of climatic conditions from the mountains of Switzerland, to the frozen north of Sweden and Finland, to the marginal coastal areas of the United Kingdom. In Europe and North America snowcasting is still perceived to be the greatest research challenge to the meteorologists, and road-surface sensors will play an

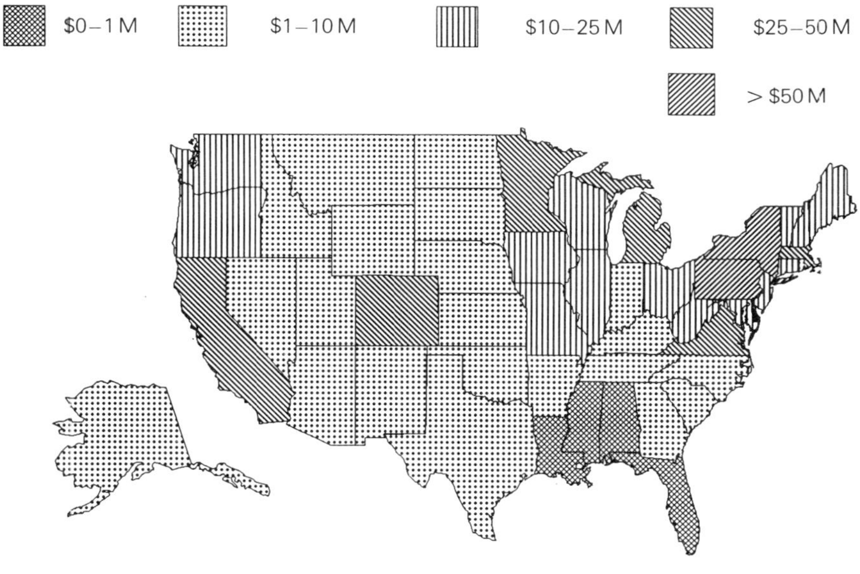

Figure 3.12 Annual snow and ice control costs ($) in the United States (M = millions).

important role, enabling up-to-date information to be fed into meso-scale prediction models. The 1990s will hopefully see this information passed on to road users.

3.8 REFERENCES

Beaumont, T.E., *et al.* (1987). An improved airborne thermal mapping technique for winter maintenance. *Mun. Engr.*, **4**, 75–86.

Bogren, J. (1990). Application of a Local Climatological Model for Prediction of Air and Road Surface Temperatures. *GUNI Report 31*, University of Gothenburg, Sweden.

Boselly, S.E. *et al.* (1990). The SHRP; addressing road weather information and applications in the US. *Proceedings of the VIth International Conference on Interactive Information and Processing for Meteorology, Oceanography and Hydrology*, Anaheim, USA, 7–14.

Department of Transport (1988a). *Specification – National Ice Prediction Network*, Ref. No. NEB 20209

Department of Transport (1988b). *Thermal Mapping Specification*, Ref. No. HM-TMI.

Gustavsson, T. (1990). Modelling of Local Climate: with applications to winter road conditions. *GUNI Report 30*, University of Gothenburg, Sweden.

Hulme, M. (1982). A new winter index and geographical variations in winter weather. *Journal of Meteorology*, **7**(73), 294–300.

McClatchy, J. *et al.* (1987). Satellite images of extremely low temperatures in the Scottish Highlands. *Met. Mag.*, **116**, 376–85.

Ministry of Communication (1982). *Road Weather Service Development*, Finland, TVH 722329, 62pp.

Moore, D.F. (1975). *The Friction of Pnuematic Tyres*, Oxford: Elsevier Scientific, 220pp.

Olafsson, I. (1985). The Swedish Experience with Road Weather Systems, In *Second International Road Weather Conference*, Copenhagen, Denmark.

Parmenter, B. and Thornes, J.E. (1986). The use of a computer model to predict the formation of ice on road surfaces. *Research Report RR 71*, TRRL, 19pp.

Perry, A. and Symons, L. (1986). Winter road sense. *Geographical Magazine*, **58**, 628–631.

Ponting, M. (1984). Weather prediction systems. *Highways and Transportation*, **31**, 24–32.

Ponting, M. (1989). *Highway Weather Forecasting During Winter*, M.Sc. Thesis, University of Birmingham, unpublished.

Rayer, P.J. (1987). The meteorological office forecast road surface temperature model. *Met. Mag.*, **116**, 180–191.

Scharsching, H. (1988). A method for detection of road surface conditions without contact. *Proceedings of the IVth International Conference on Weather and Road Safety*, Florence, pp. 183–190.

Thornes, J.E. (1985). The prediction of ice formation on roads. *Highways and Transportation*, **32**(8), 3–12.

Thornes, J.E. (1988). Ice Prediction Enters New Phase, *Highways Surveyor*, 22–24.

Thornes, J.E. (1989). A preliminary performance and benefit analysis of UK national road ice prediction system. *Met. Mag.*, **118**, 93–99.

Thornes, J.E. and Fairmainer, B. (1989). Making the correct predictions. *Surveyor*, **172**, 22–24.

Thornes, J.E., Shao, J. and Boselly, E. (1990). A new winter index and it's application in the US. *The United States Strategic Highway Research Program*, Thomas Telford, London.

Thornes, J.E. and Shao, J. (1991). A Comparison of UK Road Ice Prediction Models. *Met. Mag.*, **120**, 51–7.

Chapter Four

Snow and ice control in North America

A.H. Perry and J. Nanninga

Snow and ice control techniques in North America, and especially in the USA, have tended to focus not on preventing bond formation between ice and the road by pre-salting, but on bond destruction. Efficient clearance of highways, snow-fighting programmes, and even ploughing competitions have been the pre-occupation of municipalities, counties and states. There are probably two major reasons for this difference in emphasis as compared with that in Europe:

(a) early experiments with ice-detection sensors were disappointing, with unreliable equipment and whole systems that did not work satisfactorily;
(b) there was a lack of communication between highway engineers and meteorologists (Thornes 1986).

In addition, in many parts of the mid-West and the North-East, in what is often called the 'snow belt', large cities and conurbations have annual expectations of snow exceeding one metre, and in bad winters blizzard conditions can easily dump two or three times this amount on these settlements. For example in Wisconsin an average winter season has 30 winter storms which leave about 130 cm of snow.

4.1 SNOW PROGRAMMES

Traditionally a workable snow plan, which needs to be perceived by the general public as efficient, value-for-money and equitable, is the goal of most snow control agencies. A recent survey by Kuennen (1989) of over 400 such agencies found that a successful snow-fighting programme had three priorities:

(a) Good relations with local news media help to disseminate warnings and information to road users.
(b) A good information 'package' is needed for the general public and particularly for drivers. Pre-season snow brochures giving details of priority routes, parking bans for streets and driving tips are widely used. Newspapers and newsletters are also used to disseminate information. In Metropolitan Toronto, if 5 cm or more of snow falls in an 8 h period, a snow emergency may be declared with parking and stopping in specific streets in the downtown area prohibited.
(c) A highly trained workforce of 'snow fighters' is given regular training. Montreal, for example, employs 1400 'white knights' who undergo training and attend refresher courses in snow removal each year.

Snow and ice control is the single most costly maintenance function for many northern states and cities. Nationally the direct cost of winter maintenance activities in the US i.e. the cost of equipment, labour and materials, totals nearly $2000 million (Minsk 1989). Estimates of the losses arising from the environmental impact of the use of de-icing chemicals range from $800 million to $2000 million.

4.2 CONTROL TECHNIQUES

Snow and ice control techniques vary widely across the United States since no national direction has been mandated. At the state level there are, in effect, fifty separate authorities, each with its own operating standards and practices: cities, towns and, in many cases, counties add to the number of independent authorities following their own practices. Dissemination of information on good winter-maintenance practices is accomplished by state governmental organizations (the American Association of State Highway and Transportation Officials), federal governmental agencies (Federal Highway Administration), and private organizations (e.g. the Transportation Research Board at the National Academy of Science), as well as by industrial organizations (e.g. material and equipment manufacturers).

Salt with abrasives is by far the most popular category of de-icer, and about ten million tonnes are used in the US each year and a further three million tonnes in Canada. The influential Salt Institute has emphasized the fact that salt remains the most economical, safe and efficient de-icer, and by promoting a sensible salting programme has tried to defuse the arguments of the environmental lobby. This pro-

Table 4.1 *Stormfighting guidelines (Salt Institute)*

Condition 1 Temperature: near 30 °F Precipitation: Snow, sleet or freezing rain Road surface: Wet	If snow or sleet, apply salt at 500 lb per two-lane mile. If snow or sleet continues and accumulates, plough and salt stimultaneously. If freezing rain, apply salt at 200 lb per two-lane mile. If rain continues to freeze, re-apply salt at 200 lb per two lane-mile.
Condition 2 Temperature: Below 30 °F or falling Precipitation: Snow, sleet or freezing rain Road surface: Wet or sticky	Apply salt at 300–800 lb per two-lane mile, depending on accumulation rate. As snowfall continues and accumulates, plough and repeat salt application. If freezing rain, apply salt at 200–400 lb per two-lane mile.
Condition 3 Temperature: Below 20 °F and falling Precipitation: Dry snow Road surface: Dry	Plough as soon as possible. Do not apply salt. Continue to plough and patrol to check for wet, packed or icy spots; treat them with heavy salt applications.
Condition 4 Temperature: Below 20 °F Precipitation: Snow, sleet or freezing rain Road surface: Wet	Apply salt at 600–800 lb per two-lane mile, as required. If snow or sleet continues and accumulates, plough and salt simultaneously. If temperature starts to rise, apply salt at 500–600 lb per two-lane mile, wait for salt to react before ploughing. Continue until safe pavement is obtained.
Condition 5 Temperature: Below 10 °F Precipitation: Snow or freezing rain Road surface: Accumulation of packed snow or ice	Apply salt at rate of 800 lb per two-lane mile or salt-treated abrasives at rate of 1500–2000 lb per two-lane mile. When snow or ice becomes mealy or slushy, plough. Repeat application and ploughing as necessary.

Note: The light, 200 lb application called for in Conditions 1 and 2 must be repeated often for the duration of condition.
(30 °F = −1 °C; 20 °F = −6.7 °C; 10 °F = −12.2 °C; 100 lb = 45 kg)

Table 4.2 *Sample form provided by private weather service.*

STORM WARNING SERVICE

PROVIDED BY

AIR SCIENCE CONSULTANTS, INC.

PHONE
412 - 221-6002

347 PRESTLEY ROAD
BRIDGEVILLE, PA 15017

Storm No. ____________

Report No. ____________

Date ____________

ALERT

Time_____EST Date_____19___

Issued by: ____________

Received by: ____________

DEPENDABILITY OF ALERT:

1. Excellent ☐ 3. Fair
2. Good

STORM CONDITIONS:

4. Freezing Rain
5. Snow (Less than 2 inches)
6. Snow (2 inches to 4 inches)
7. Snow (4 inches or more)

STORM EXPECTED TO BEGIN:

8. During rush hour 6-9 a.m.
9. During rush hour 4-7 p.m.
10. Within 6-12 hours
11. Within 12-24 hours
12. Within 24-36 hours
13. Within 36-48 hours
14. Within 48-72 hours

G. ENDING OF MAIN STORM:

☐ 47. Storm Ending_____ and_____, day_____.
☐ 48. Storm ending time uncertain

H. SNOW TYPE:

☐ 49. Dry (Below 30° F)
☐ 50. Wet (30-34° F)
☐ 51. Melting (over 34° F)
☐ 52. Wet becoming dry
☐ 53. Dry becoming wet

L. SNOWFALL AMOUNT-MAINSTORM:

☐ 54. Less than 2 inches
☐ 55. 1-3 inches ☐ 59. 6-9 inches
☐ 56. 2-4 inches ☐ 60. 8-12 inches
☐ 57. 3-5 inches ☐ 61. 10-15 in.
☐ 58. 4-7 inches ☐ 62. 12-18 in.
☐ 63. Over 18 inches
☐ 64. Amount Excellent
☐ 65. Amount Good

J. MAIN SNOW ACCUMULATION RATE:

☐ 66. First 2 inches between _____ and _____, day _____.
☐ 67. _____ inches between _____ and _____ day _____.

☐ 99. Air temp. remaining above 32° F
☐ 100. Air temp. hovering near 32° F
☐ 101. Air temp. rising or steady
☐ 102. Air temp. falling or steady
☐ 103. Pavement warmer than temp.
☐ 104. Pavement same as air temp.
☐ 105. Pavement colder than air temp.

O. WIND DETAILS:

☐ 106. Wind light and variable
☐ 107. N. ☐ 111. N.
☐ 108. E. ☐ 112. E.
☐ 109. S. ☐ 113. S.
☐ 110. W. ☐ Becoming ☐ 114. W.
☐ 115. Time of windshift _____ day between _____ and _____
☐ 116. Wind 10-15 mph.
☐ 117. Wind 15-25 mph.
☐ 118. Wind 25-40 mph.
☐ 119. Wind over 40 mph.
☐ 120. Wind increasing during storm
☐ 121. Drifting winds during storm

P. SNOWSHOWERS:

☐ 122. Occasional flurries
☐ 123. Light snowshowers

R. SPECIAL ROAD PROBLEMS:

☐ 152. Rush hour trouble will occur ☐ a.m. ☐ p.m. ☐ today ☐ tomorrow
☐ 153. Snow packing
☐ 154. Drifting
☐ 155. Rapid accumulation of snow
☐ 156. Wet snow or slush freezing
☐ 157. Icing at intersections
☐ 158. Local icing - viaducts, bridges
☐ 159. Alert crews for possible trouble
☐ 160. Spread salt or abrasives
☐ 161. Get salt down early
☐ 162. General plowing condition

S. WEATHER CONDITIONS AFTER MAIN STORM:

☐ 163. Cold wave (temps. below 10° F)
☐ 164. Below freezing temps.
☐ 165. Slowly rising temps.
☐ 166. Rapid thaw
☐ 167. Thawing conditions within 12 hrs.
☐ 168. Thawing conditions within 24 hrs.
☐ 169. Above freezing days - freezing nights
☐ 170. Drifting winds (25-40 mph.)
☐ 171. Heavy drifting winds (over

REVISION COLUMN STORM WARNING DETAILS

U. REVISION COLUMN - MAIN STORM

Beginning time:

Date Time Issued by Rec. by

☐ 181. Storm beginning _____ and _____, day _____.

Ending Time:

☐ 182. Storm ending _____ and _____, day _____.

V. SNOWFALL AMOUNT - MAIN STORM:

☐ 183. Less than 1 inch
☐ 184. Less than 2 inches
☐ 185. 1-3 inches
☐ 186. 2-4 inches
☐ 187. 3-5 inches
☐ 188. 4-7 inches
☐ 189. 6-9 inches
☐ 190. 8-12 inches
☐ 191. 10-15 inches
☐ 192. 12-18 inches
☐ 193. Over 18 inches

STORM WARNING DETAILS

Time_____EST Date_____19___

Issued by: ____________

Received by: ____________

DEPENDABILITY OF DETAILS:

- 15. Excellent ☐ 17. Fair
- 16. Good

TYPE OF TROUBLE - MAIN STORM:

- 18. Snow
- 19. Snowshowers (See sec. P)
- 20. Freezing Rain, Freezing Drizzle, Sleet
- 21. Rain changing to Freezing Rain
- 22. Freezing rain changing to Snow
- 23. Rain changing to Snow
- 24. Freezing rain changing to Rain
- 25. Snow changing to Freezing Rain
- 26. Snow changing to Rain

MAIN STORM - BEGINNING TIME:

- 27. Timing Excellent
- 28. Timing Good
- 29. 12-2 ☐ 39. 10-12
- 30. 1-3 ☐ 40. 11 a.m.-1 p.m.
- 31. 2-4 ☐ 41. 11 p.m.-1 a.m.
- 32. 3-5 ☐ 42. a.m.
- 33. 4-6 ☐ 43. p.m.
- 34. 5-7 ☐ 44. Today
- 35. 6-8 ☐ 45. Tonight
- 36. 7-9 ☐ 46. Tomorrow
- 37. 8-10
- 38. 9-11

☐ 68. _____ inches between _____ and _____ day _____.

K. RAIN EXPECTED

- ☐ 69. 12-2 ☐ 78. 9-11
- ☐ 70. 1-3 ☐ 79. 10-12
- ☐ 71. 2-4 ☐ 80. 11 a.m.-1 p.m.
- ☐ 72. 3-5 ☐ 81. 11 p.m.-1 a.m.
- ☐ 73. 4-6 ☐ 82. a.m.
- ☐ 74. 5-7 ☐ 83. p.m.
- ☐ 75. 6-8 ☐ 84. Today
- ☐ 76. 7-9 ☐ 85. Tonight
- ☐ 77. 8-10 ☐ 86. Tomorrow

L. RAIN TYPE:

- ☐ 87. Hard Freezing (Below 25° F)
- ☐ 88. Freezing (25-32° F)
- ☐ 89. Cold (32-38° F)

M. EFFECT OF RAIN:

- ☐ 90. Turn snow to slush
- ☐ 91. Wash out snow

N. TEMPERATURE DETAILS:

- ☐ 92. Air temp. dropping *below* 32° F _____ and _____ day _____
- ☐ 93. Air temp. dropping *below* 25° F _____ and _____ day _____
- ☐ 94. Air temp. dropping *below* 15° F _____ and _____ day _____
- ☐ 95. Air temp. rising *above* 25° F _____ and _____ day _____
- ☐ 96. Air temp. rising *above* 32° F _____ and _____ day _____
- ☐ 97. Air temp. remaining below 32° F
- ☐ 98. Air temp. remaining below 25° F

- ☐ 124. Moderate snowshowers
- ☐ 125. Heavy snowshowers
- ☐ 126. Beginning within 6 hrs.
- ☐ 127. Beginning within 6-12 hrs.
- ☐ 128. Beginning within 12-18 hrs.
- ☐ 129. Beginning within 18-24 hrs.
- ☐ 130. Beginning after 24 hrs.
- ☐ 131. Lasting less than 6 hrs.
- ☐ 132. Lasting 6-12 hrs.
- ☐ 133. Lasting 12-24 hrs.
- ☐ 134. Lasting 24-48 hrs.
- ☐ 135. Lasting more than 48 hrs.
- ☐ 136. Showers lasting indefinitely
- ☐ 137. Less than ½ inch accumulation
- ☐ 138. Accumulation ½-1 inch
- ☐ 139. 1-2 inches
- ☐ 140. 2-4 inches
- ☐ 141. Snowshower acc. in addition to main storm trouble

Q. EXPECTED ROAD CONDITIONS

- ☐ 142. Main roads will be slick ☐ tonight ☐ early a.m. ☐ all day
- ☐ 143. Secondary roads will be slick ☐ tonight ☐ early a.m. ☐ all day
- ☐ 144. Main roads will be wet
- ☐ 145. Main roads will be snowcovered
- ☐ 146. Main roads will be snowcovered in spots
- ☐ 147. Secondary roads will be wet
- ☐ 148. Secondary roads will be snowcovered
- ☐ 149. Secondary roads will be snowcovered in spots
- ☐ 150. Roads will be ice covered
- ☐ 151. Roads will be ice covered in spots

40 mph.)

- ☐ 172. Winds decreasing within 6 hrs.
- ☐ 173. Winds decreasing in 6-12 hrs.
- ☐ 174. Winds decreasing in 12-18 hrs.
- ☐ 175. Winds decreasing in 18-24 hrs.
- ☐ 176. Winds decreasing after 24 hrs.
- ☐ 177. New storms on way - details later

T. COMMENTS:

- ☐ 178. Storm dissipating; cancel warning
- ☐ 179. No road problems expected
- ☐ 180. Remarks: ____________

____

____

____

____

____

____

____

____

____

____

____

☐ 194. Snow will accumulate 2 inches between _____ and _____ day _____.

- ☐ 195. Amount Excellent
- ☐ 196. Amount Good

W. SNOWSHOWERS:

____

Date Time Issued by Rec. by

- ☐ 197. Showers beginning within 6 hrs.
- ☐ 198. Showers beginning within 6-12 hrs.
- ☐ 199. Showers beginning within 12-18 hrs.
- ☐ 200. Showers beginning within 18-24 hrs.
- ☐ 201. Showers beginning after 24 hrs.
- ☐ 202. Showers lasting 6-12 hrs.
- ☐ 203. Showers lasting 12-24 hrs.
- ☐ 204. Showers lasting more than 24 hrs.
- ☐ 205. Little or no accumulation
- ☐ 206. Less than 1 inch
- ☐ 207. 1-2 inches
- ☐ 208. 2-4 inches
- ☐ 209. 3-5 inches

gramme emphasizes the need for good salt storage, the calibration of spreaders, and trained personnel to apply salt to roads. Table 4.1 shows suggested 'storm-fighting guidelines' that should be employed with different combinations of meteorological conditions. Historically, salt has been mixed with abrasives to reduce the overall costs of material, although this can lead to increased clean-up costs in Spring. Butte, Montana mounted a campaign for straight salting due to the dust and grime created by sanding city streets. Advocates of salting point to the experience of four Massachusetts towns that banned salt for varying periods. One of them, Concord, lifted its ban in mid-winter after experiencing a 30% increase in accidents.

Salt substitutes have been tried in a number of areas as de-icers, especially liquid calcium chloride, but 'exotic' alternatives to salt remain little used. Corrosion-inhibiting methods are now being used in building new bridges and protecting existing ones. Bridge decks can be protected with cathodic protection, where a small reverse current halts the rusting process, while new bridges use epoxy-coated reinforcing rods for protection. Air-entrained concrete or high-density concrete can also prevent concrete spalling.

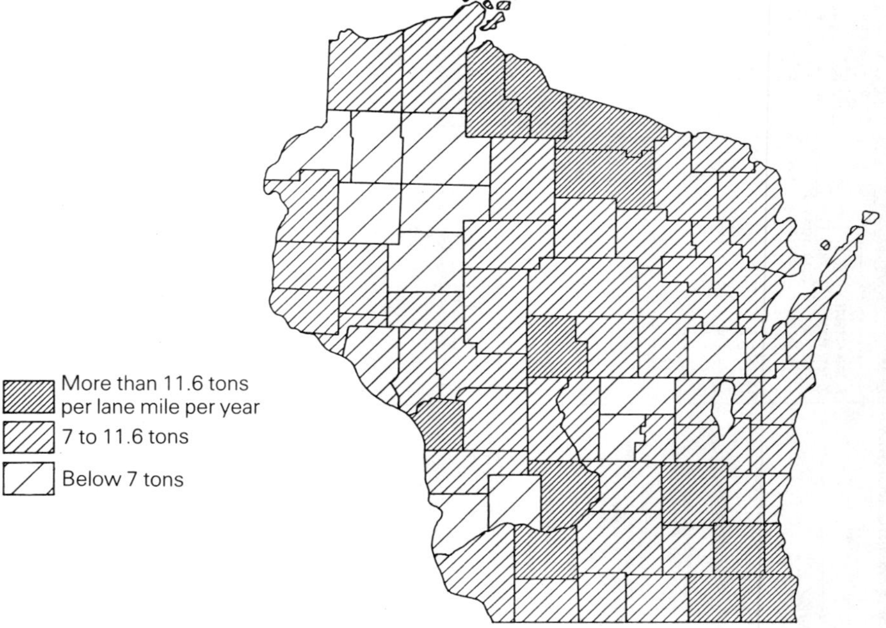

Figure 4.1 Variation in salt usage per county in Wisconsin, averaged over three recent winters. Amounts in US tons.

It is hard to avoid the conclusion that the use of salt in conjunction with a good ploughing programme is the fastest and most efficient means of snow and ice removal. The use of abrasives requires at least seven times more material to treat a given distance of roadway. Salt is likely to remain an essential part of the winter environment, accepted and tolerated by the motoring public, despite its corrosive powers. It is plentiful, safe and easy to handle and non-toxic to man, and its benefits in use outweigh its detrimental effects at a ratio of about 18 to 1. Salt usage in US tons per total lane mile, varies from 10–25 tons around the Great Lakes to less than one ton in many western states.

If snow removal is to be carried out efficiently then accurate weather forecasts are essential. Many authorities in addition to making use of the National Weather Service, have also contracted private weather services. Such services can provide specific forecasts in a format that the client requires, and an example of the kind of detailed service that can be provided is illustrated in Table 4.2. Snow and ice removal is achieved more easily, at less cost and with less disruption to drivers, if the proper remedial treatment is initiated at the start of a storm, and this is where good forecasting can pay for itself.

4.3 WISCONSIN: A CASE STUDY

Wisconsin sits in the middle of the so called 'rust belt', and uses about half a million tons of salt each winter: 5% of the national total. The spatial variation in salt usage per county is shown in Figure 4.1 expressed in tons per lane mile averaged over three recent winters. Variation in spreading rates are considerable, probably because traditionally each county has had considerable autonomy and salting rates have become fossilized (Thornes 1986). A winter highway classification has been operative in Wisconsin since 1973/74 which creates three classes of highway, each of which has a different policy for snow ploughing:

(a) *Class 1.* Roads carrying 5000 vehicles or more daily. These roads will have a 24 h snow-ploughing and ice-control service seven days per week. About 2500 miles (4000 km) of state highway are in this category, out of a total of 108 000 miles (174 000 km) of road in the state.
(b) *Class 2.* Roads carrying 1–5000 vehicles daily should have bare surfaces as soon as practical but reduced services operate at night on these roads.
(c) *Class 3.* Roads carrying less than 1000 vehicles daily should have single lane bare carriageways as soon as practical, but such roads receive no service at night.

Wisconsin, like a number of other states, has recently begun to follow a more 'European' approach to its winter maintenance. Thermal mapping has been started and 25 remote weather stations with electronic sensing devices have been installed (Stephenson 1989), and information from this network is used to assist in the performance of winter highway operations. In Wisconsin the National Weather Service has only two stations from which to make its forecasts and the installed network of sensors is seen as a way of increasing the accuracy of weather information.

4.4 COLLABORATION

The American Public Works Association through its annual conference, local chapters (centres) and publications is the principal forum for trading ideas and suggestions with managers from almost every snowbelt state and almost all Canadian provinces.

4.5 THE STRATEGIC HIGHWAY RESEARCH PROGRAMME

SHRP is a five-year research programme established by the US National Academy of Sciences. Snow and ice control were identified at an early stage as areas of study. In particular the mechanics of the ice-road bond will be investigated with one study from the viewpoint of preventing the bond from forming and the other concentrating on destroying the bond after it has formed. By the time the project concludes in 1993 SHRP expects to have demonstrated how road surface modifications can reduce ice adhesion. Other studies will evaluate de-icing chemicals and study how improved storm warning and communication capabilities will aid highway authorities. All studies will rank the potential pay-offs, costs, and cost-effectiveness of various approaches and validate the best of them by field trials.

4.6 REFERENCES

Kuennen, T. (1989). Snow budgets stable, PR efforts up, survey show. *Roads & Bridges*, **27**, 40–47.

Minsk, D. (1989). SHRP snow, ice contracts may aid snow fighters. Roads & Bridges, **27**, 48.

Stephenson, T.E. (1989). Wisconsin's winter weather system. *Proc. 4th Intern. Conf. on Weather & Road Safety, Florence*, pp. 61–102.

Thornes, J. (1986). Snow and ice control in N. America. *Highway Meteorology*, **2**, 8–18.

Chapter Five

Snow-drift modelling and control

S.L. Ring

In cold regions, blowing snow may create hazardous conditions for travellers. In certain terrain, roads may become impassable and even with the most modern highway design some vision problems and snow-drift accumulation may occur. This is a distinct problem for highway engineers as today's drivers expect a clear roadway, and object to the delays and unsafe conditions associated with blowing snow.

Modern technology has provided the design capability for and construction of highway facilities relatively free of snow-drifts. Regrettably not all highways can economically incorporate these concepts due to terrain and other restrictions. Also, some highway design and maintenance personnel are not familiar with the technical concepts involved and hence are part of the problem. In fact, snow-drifts are created through this lack of fundamental knowledge of transport in snow conditions which results, for example, in snow fences at the wrong location.

Control of wind-blown snow by the highway engineer is usually categorized in one of three areas:

(a) shaping the roadway on the topography such that an aerodynamically clean cross-section results;
(b) strategically placing barriers, such as snow fences or plantings, which will create the snow-drifts off the highway and thus result in smaller snow quantities available for deposition on the highway;
(c) removing inappropriate barriers or spot-improving the cross-section so that a snow-drift-prone location may be improved.

5.1 FUNDAMENTAL CHARACTERISTICS OF WIND-BLOWN SNOW

Understanding the characteristics of wind-blown snow is basic to the control of drifting snow. However, the movement of small, loose particulates in the wind is a complex phenomenon. Particles can move across the environment in one of three ways: by creep, saltation, or suspension. *Creep* is the motion in which large particles do not become airborne but just roll along the surface. *Saltation* occurs when the particle rises from the surface in a nearly vertical direction and then gradually returns via a shallow angle. *Suspension* occurs when the particle stays airborne due to small size and excessive wind velocity.

Nearly all blowing snow particles move in the saltating mode (Kind 1976; Ring *et al.* 1979); the threshold wind speed associated with saltating snow movement under average conditions of a flat, smooth field, covered with dry snow, would be approximately 8–35 m/s (Tabler 1978).

Snow is a cohesive material and particles tend to cohere together easily. When cold and dry the particles collect an electrostatic charge when blown across a surface and may have an angle of repose greater than 90°. In fact this cohesive feature results in cornices and overhangs as the drift develops.

Another phenomenon of blowing snow is that snow particles do not move over long distances. When blown by strong winds, snow crystals are broken and abraded, and through the process of sublimation, they disappear. The result is a finite quantity of snow in a given area.

Snow-drift configurations have been studied extensively for years and the shape of the equilibrium final drift profile is predictable. The characteristics of the drift-creating barrier dictate the resultant shape of the final drift for a given quantity of snow. A major research effort has been concerned with the shape of equilibrium snow-drift profiles for establishing size, shape, and location of snow fences.

Concepts in snow fence design resulting from these research studies are presented later in this chapter.

5.2 SNOW-DRIFT RESEARCH

A considerable amount of research effort has been concerned with snow transport and snow-drifting. Many of these studies are directly applicable to highway design and maintenance and the results have provided design standards and criteria for those concerned with maintaining clear roadways in cold regions.

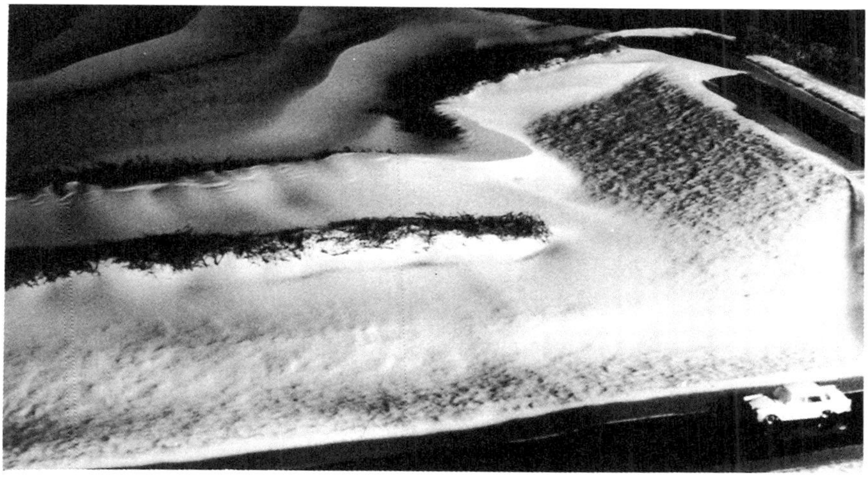

Figure 5.1 Wind-tunnel modelling of snow-drifts created by plantings.

Some of the first modelling experiments in a wind tunnel were conducted by Finney (1934) at Iowa State University. Wind-tunnel testing of snow-fence models includes studies by Becker (1944), Nokkentved (1940), and Finney (1937). Considerable modelling in wind tunnels has been performed at Iowa State University and at the NASA Ames Research Center in the United States in recent years, by Iverson *et al.* (1973, 1975, 1976a), and by Greeley *et al.* (1974) on drifting sand and dust. Probably the most definitive study on wind-tunnel modelling of a living snow fence (plantings) for highways is by Ring *et al.* (1979). Figure 5.1 illustrates the wind-tunnel analysis of snow-drifts created by plantings. The 1979 report by Ring *et al.* (1979) covers in detail the similitude aspects of wind-tunnel modelling and the various media suitable for use in a wind tunnel experiment. An extensive bibliography is included.

In recent years Tabler's work (1973, 1974, 1975, 1978, 1979) can be considered classic in snow-fence design. Tabler did not use the wind tunnel, but an actual field environment. Some of his experiments were full-scale observations and some were reduced-scale on ice areas.

In addition to wind-tunnel modelling and natural environment experiments, Theakston (1970) has conducted numerous experiments in water. Also, de Kransinski and Anson (1975) have modelled accumulation of snow-drifts around buildings using water flumes. However, questions have been raised regarding sediment transport in water as an appropriate modelling medium for snow in air (Ring *et al.* 1979).

It should be noted that applying snow-drift mathematical modelling concepts can have complications. Unusual terrain, adjacent buildings or bands of plantings, steep cuts or embankments, fences and bridges are all examples of unique conditions making it difficult to predict where snow will be deposited. Under certain conditions wind currents may be difficult to predict and the only way to offer guidance is through the use of modelling techniques.

5.3 THE SNOW-FREE HIGHWAY DESIGN CONCEPT

Highway designers have recognized the relationship between roadway cross-section and the propensity for snow-drifting in particular areas. Even the novice maintenance worker soon recognizes that an elevated grade above the surrounding land and a cross-section with flat slopes will generally remain clear of snow-drifts. In fact some state agencies in the United States have developed sophisticated computer programs with interactive graphics to evaluate snow-drift-prone locations.

Finney (1934, 1937, 1939) published the results of numerous experiments in a wind tunnel concerning the control of drifting snow through the design of the roadway. Finney's *Bulletin Number 86* (1939) was a classic work, and in recent years Tabler's publications (1973, 1974, 1975, 1978) have provided new insight into the snow-drifting phenomenon and especially snow-fence design. Ring's 1979 publication was concerned with living snow fencing.

The primary consideration in the design of a snow-drift-free roadway cross-section is to keep the snow moving. When air pressure changes occur, eddy areas develop causing the snow to drop and form drifts. Early studies by Finney (1934a) determined that the limits of the eddy area from a solid wall type of snow fence (e.g. the relatively steep backslope from the highway ditch up to the adjacent land) would be

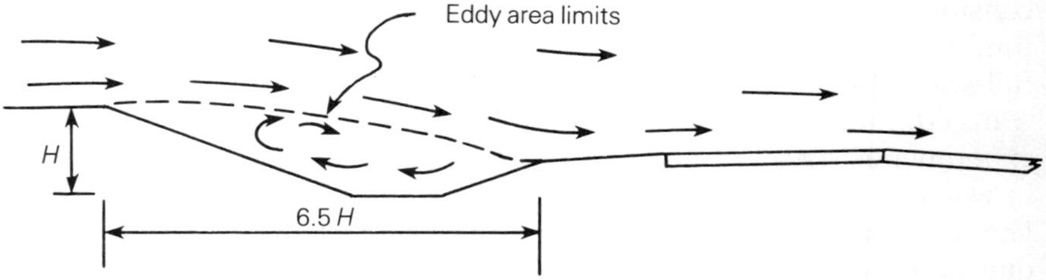

Figure 5.2 Drift forming eddy in a highway cut (Finney 1934a).

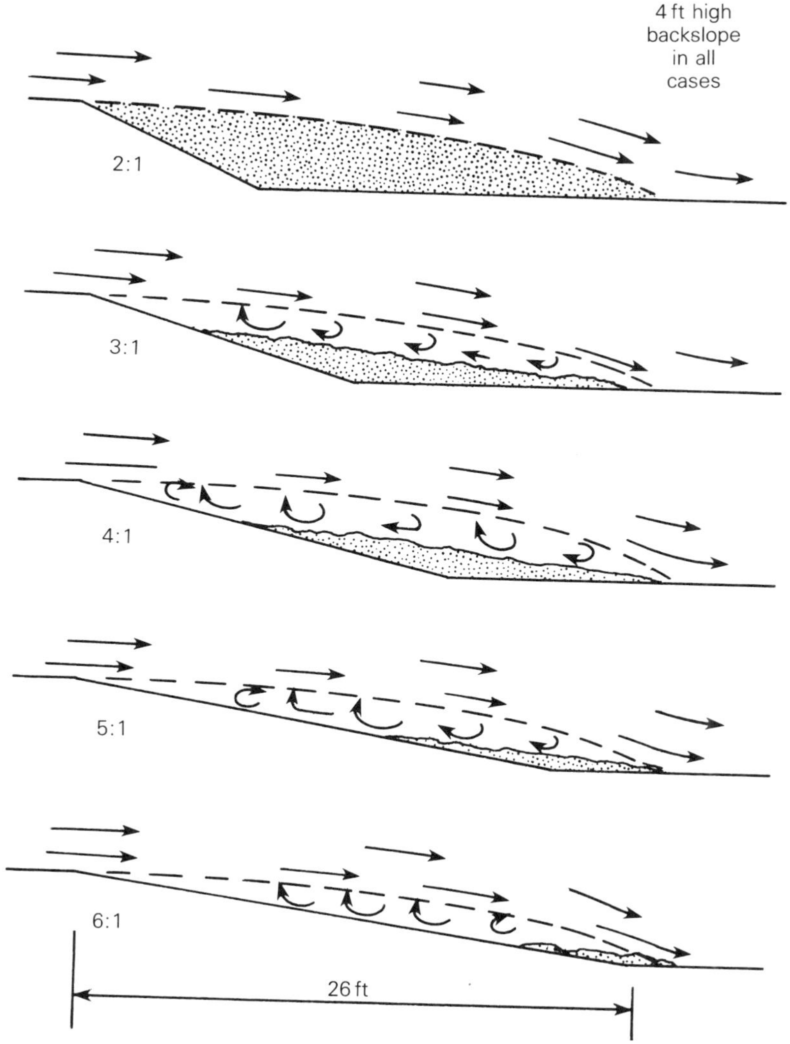

Figure 5.3 Effect of variations in backslope on snowdrift accumulation (Finney 1934a). (1 ft = 0.305 m).

6.5 *H*, where *H* is the height of obstruction, or in this case the depth of cut from adjacent land to the ditch bottom (see Figure 5.2).

This characteristic is a powerful tool in predicting snow-drift length (versus the height of fence *H*) and can also be used for locating solid snow fences. Refinements since these 1930s tests by Tabler (1975)

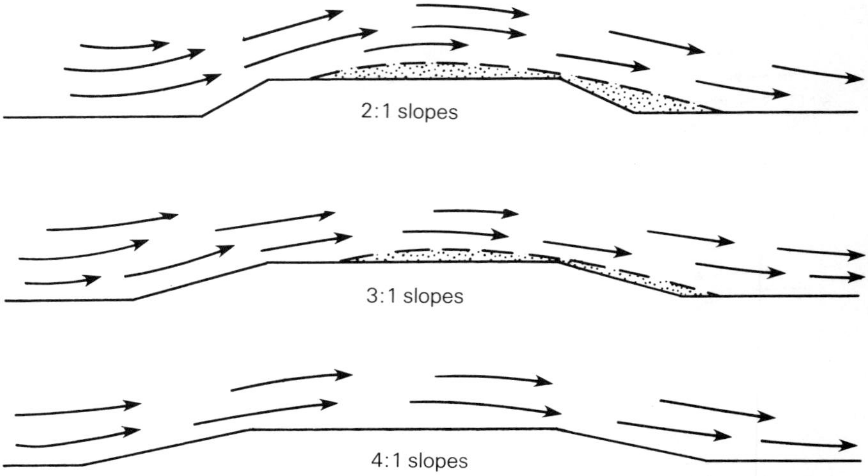

Figure 5.4 Effect of foreslopes on air flow across roadways (Finney 1934a).

have shown that a solid barrier (functioning similar to a cut section's backslope) may generate drift lengths as much as 30 H for cuts of 5 ft (1.5 m) or less and may be as low as 7 H for the deeper cuts, greater than 10 ft (3 m) (Figure 5.3). These early studies have also shown that flat foreslopes on the roadway can be self-cleaning. Figure 5.4 illustrates the concept.

Modelling studies have shown that a 4 : 1 foreslope is relatively aerodynamically clean and a 6 : 1 slope provides an excellent self-cleaning cross-section. Finney (1939) made the following recommendations which still hold true today:

(a) Raise the grade line above the adjacent ground equal to the average depth of snow accumulation.
(b) Avoid cut sections through alignment and profile design. Shallow-cut sections (less than 2 m) are troublesome.
(c) When cut sections are unavoidable use concepts of snow transport and storage in the design of the cross-section (e.g. wider ditches, flatter backslopes, and rounded-slope intersections).
(d) In general, use flat slopes (4 : 1 or greater), wide shoulders, and shallow, wide ditches.
(e) Eliminate guard rail if possible and other appurtenances such as kerbs.
(f) Utilize a knowledge of snow transport phenomena in highway location.

Figure 5.5 Smooth cross-section and flat foreslopes provide drift-free roadway.

Cron (1967) called attention to the problems of information on prevailing wind and topographic conditions in highway location and design. He also noted that appurtenances such as kerbs, guard rails, signs and fences function as snow fences.

Tabler's 1975 publication 'Predicting profiles of snowdrifts in topographic catchments' is representative of the value of his work and many contributions to the state of the art. He has verified some of the earlier research results, and provided much new information for the highway designer. Using these concepts for designing a snow-drift-free highway cross-section, where economically feasible, the designer will reduce future maintenance and operational problems. Figure 5.5 illustrates this type of cross-section. It is important that all appurtenances that could function as snow fences be eliminated if possible from the design. This is especially true of bridge piers that will require guard rails, kerbs and signs.

5.4 THE DESIGN OF SNOW FENCES AND OTHER BARRIERS

Those persons concerned with street and highway design, maintenance, or operations should have a knowledge of the fundamentals of the snow-drift phenomenon. In many cases they do in fact have an intuitive judgement developed from experience, or they operate from policies developed from field observations that have served them well in past situations.

There are three primary types of snow fences according to Martinelli (1973) and Pugh and Price (1954): the *collecting fence*, which is a solid or porous barrier that decelerates the wind speed providing snow deposition; the *solid guide fence*, which is aligned at an angle to the wind (in plan) in order to deflect the snow laterally; and the *blower fence*, which is aligned at an angle to the wind (in elevation) in order to accelerate the wind speed locally and cause the snow to be transported elsewhere.

By far the most usual situation calls for a collecting fence. It is used upwind and adjacent to the highway in order to collect and deposit snow before it reaches the roadway. In general these fences are parallel to the highway where the wind is normal to the highway; or short sections of staggered fences may be used to achieve a right-angle condition where the wind is quartered to the highway.

In practice, the usual situation calls for predicting the final snow-drift shape which would result from a particular snow fence placed upwind of the highway, so that the resulting drift will not encroach on the highway. The maximum snow-drift accumulation occurs at equilibrium conditions where any additional snow will blow across the drift and not be deposited. Intensive research using actual field conditions as well as wind-tunnel modelling has been conducted and the results and recommendations regarding snowdrift profiles are well documented. Ring *et al.* (1979) has an extensive literature review and overview of the state of the art regarding the snow-drift profiles for various types of snow fence.

On a flat landscape with no obstructions, such as trees or buildings, the snow-drift profile resulting from a particular snow fence is predictable. However, it should be noted that there is some minor disagreement among the authorities regarding details of snow-drift profile dimension, but this author considers these insignificant for most applications in the highway field. The data presented herein can be used to predict the drift geometry one can expect for a given type of fence.

When the terrain is not flat, or the wind currents are erratic from trees or buildings, predicting the resultant snow-drift profile becomes more complex. In fact, for a critical complex situation, modelling may be justified to determine the expected final profile.

A fence of solid cross-section, although not generally as applicable to the highway field, develops a distinctly different drift profile from the traditional slatted snow fence. The slatted fence has openings that are

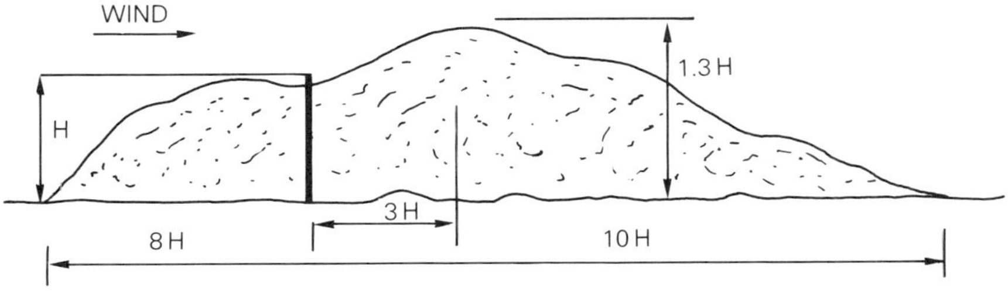

Figure 5.6 A solid-fence snow-drift profile at maximum accumulation.

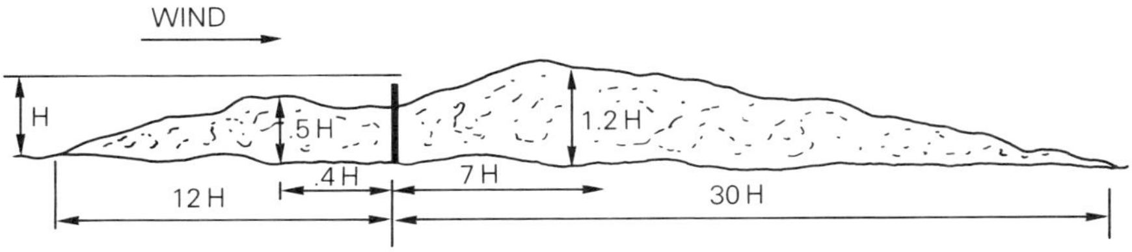

Figure 5.7 A porous-fence snow-drift profile at maximum accumulation.

about equal in area to the slats. In fact it has been determined that the most efficient fence has about a 50% porosity.

A solid snow fence will develop a leeward drift of the order of ten times the fence height (10 *H*). The maximum windward drift length will reach approximately eight times the fence height (8 *H*). These values are graphically presented in Figure 5.6. A porous fence of approximately 50% porosity will develop a leeward drift length approximately equal to 30 times the height of the fence (30 *H*), and the windward length would be 12 times the fence height (12 *H*). These values are presented in Figure 5.7.

Experts have also noted that an optimum fence geometry would have a bottom gap of about 30 cms, horizontal slats with 50% porosity, be inclined downwind at 15° from vertical and be at least 30 *H* in length. In addition a downward ground slope on the leeward side of more than 10% increases storage capacity as well as an upward ground slope on the windward side. When gaps appear in fences they should overlap at least eight times the height (8 *H*) in order to maintain maximum snow-drift accumulations as presented in Figure 5.8.

Living snow fences, such as shrubs and trees, are in use on highways in cold regions. From the results of numerous research projects it is known that height, width, porosity, and arrangement of plants

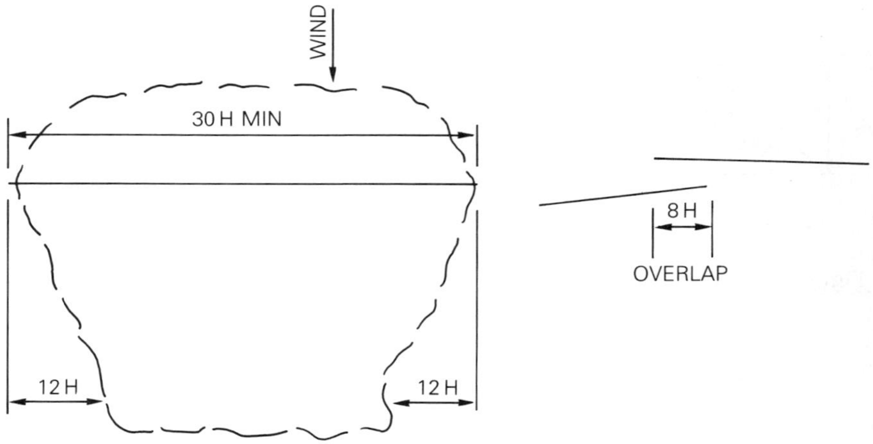

Figure 5.8 Minimum length of snow fence and overlaps.

Figure 5.9 Deep drafts at a living snow fence.

within a plant mass are the chief factors influencing snow drifting. When individual plants are grouped together in a large enough mass the potential for a significant snow storage exists. Their characteristics taken as a whole determine snow-drift configuration. The most commonly used living snow fence consists of trees and shrubs in combination in order to develop both height and porosity which will maximize the drift volume. Figure 5.9 illusrates a drift formed by a living snow fence.

Although there is a considerable amount of research on snow-drifts resulting from plants and trees, it is more difficult to arrive at definitive forecasting values. This is primarily due to difficulty in quantifying the porosity and the effective height. Even the porosity can change as wind velocity increases.

Predicting the snowdrift profile from a plant mass can be based on the concepts presented earlier. The final value predicted will probably range between that predicted for a porous fence and that for a solid fence depending on the individual's interpretation of porosity and effective height.

5.5 REMOVING INAPPROPRIATE SNOW-DRIFT-CREATING FEATURES ON THE ROADSIDE

Those persons supervising highway maintenance operations are in a position to observe snow-drift encroachment problems. Frequently these snow-drifts are the result of inappropriate plantings or weed-filled fences that are too close to the highway. In some cases the problem is created by guard rails, kerbs or other appurtenances at the edge of the roadway. As snow accumulates and is removed by a snow plough the problem is compounded by creating a vertical bank which functions as a snow fence. Figure 5.10 illustrates a typical problem. Snow fences upwind may be the only solution at this particular location.

The solution to some of these problems is to remove the obstruction that functions as a snow fence. Weed-filled fences can be cleaned, plantings close to the roadway may be removed or moved, and in some cases the guard rail or kerb may be moved.

Another problem is the ditch of inadequate width that does not provide for snow-drift storage. This is especially true near the beginning and ending of cut sections. In some cases a maintenance expenditure to flatten the backslope and/or widen the ditch will result in reduced costs of snow removal. The criteria previously presented for solid snow fences can be used to analyse these backslope problems.

Figure 5.10 Adjacent appurtenances create unwanted snow drifts on the roadway.

5.6 CONCLUSIONS

Blowing and drifting snow can create problems for travellers and can effectively shut down highway travel. Those concerned with managing modern highway systems have at their command three aspects of controlling drifting snow: designing the highway so that an aerodynamically clean cross-section is self-clearing; trapping the snow at designed snow fences upwind of the highway; or using modern snow-removal equipment and techniques to remove the snow from the highway after the drifts have formed.

Frequently the terrain involved, or the economics, will not allow a snow-drift-clear design, and snow-drifts occur on the roadway. Too often the only alternative may be to use snow-clearing equipment, or to analyse the possibility of snowfences.

Highway designers and maintenance workers have available the results of years of field observations and research based on modelling. Individuals who understand the phenomenon of snow transport are in a position to provide a snow-free highway.

5.7 REFERENCES

Becker, A. (1944). Natural snow fences along roads. *Bautechnik*, **22**, 37–42.

Cron, F.W. (1967). Snowdrift control through highway design. *Public Roads*, **34** (11), 227–234.

de Kransinski, J.S. and Anson, W.A. (1975). A study of snowdrifts around the Canada Building in Calgary. University of Calgary, Dept of Mechanical Engineering Report No. 71, October.

Finney, E.A. (1934a). Snow control on the highways. Thesis, Iowa State College.

Finney, E.A. (1934b). Snow control on the highways. *Bulletin No. 57*, Michigan Engineering Experiment Station, East Lansing, Michigan.

Finney, E.A. (1937). Snow control by tree planting, part VI: wind tunnel experiments on tree plantings. *Bulletin No. 75*, Michigan Engineering Experiment Station, East Lansing, Michigan.

Finney, E.A. (1939). Snowdrift control by highway design. *Bull. No. 86*, Michigan Engineering Experiment Station, East Lansing, Michign 1–58.

Greeley, R. *et al.* (1974). Wind tunnel studies of Martian eolian processes. *Proc. Roy. Soc. London A*, **341**, 331–360.

Iversen, J.D. *et al.* (1973). Simulation of Martian eolian phenomena in the atmospheric wind tunnel. *Space Simulation*, NASA Special Publication **336**, 191–213.

Iversen, J.D. *et al.* (1975). Eolian erosion on the Martian surface: Part 1, erosion rate similitude. *Icarus*, **26**, 321–331.

Iversen, J.D. *et al.* (1976a). Saltation threshold on Mars: the effect of interparticle force, surface roughness, and low atmospheric density. *Icarus*, **29**, 381–393.

Iversen, J.D. *et al.* (1976b). Windblown dust on Earth, Mars, and Venus. *J. Atmos. Sci.*, **33**, 2425–2429.

Iversen, J.D., *et al.* (1976c). The effect of vertical distortion in the modelling of sedimentation phenomena: Martian crater wake streaks. *J. Geophys. Res.*, **81**, 4846–4856.

Kind, R.J. (1976). A critical examination of the requirements for model simulation of wind-induced erosion/deposition phenomena such as snow drifting. *Atmos. Environ.*, **10**, 219–227.

Martinelli, M. (1973). Snow fences for influencing snow accumulation, *Proceedings on measurement and forecasting*, Banff, Alberta pp. 1394–1398.

Nokkentved, C. (1940). Drift formation at snow fences. *Stadsog Haveingenoren*, **31**(9), 111–114.

Pugh, H.L.D. and Price, W.I.J. (1954). Snow drifting and the use of snow fences. *Polar Rec.*, **7**, 4–23.

Ring, S.L., *et al.* (1979). Wind tunnel analysis of the effects of planting at highway grade separation structures. Iowa Highway Research Board Report HR202, Iowa Department of Transportation.

Tabler, R.D. (1973). New snow fence design controls drifts, improves visibility, reduces road ice. *Proc. Ann. Trans. Eng. Conf.*, **46**, 16–27.

Tabler, R.D. (1974). New engineering criteria for snow fence systems. Transportation Research Board (National Research Council), *Trans. Res. Rec.*, **506**, 65–78.

Tabler, R.D. (1975). Predicting profiles of snowdrifts in topographic catchments, Western Snow Conference, *Proceedings* **43**, 87–97, Coronado, California, April.

Tabler, R.D. (1978). Personal communication to the author during a lecture/

discussion at Iowa State University, 7 December 1978.

Tabler, R.D. (1979). Geometry and density of drifts formed by snow fences. Abstract for Snow in Motion Conference, Fort Collins, Colorado, August.

Theakston, F.H. (1970). Model technique for controlling snow on roads and runways. In *Snow Removal and Ice Control*, Highway Research Board Spec. Rep. 115, pp. 226–230.

Chapter Six

The fog hazard

Leslie F. Musk

The main meteorological hazards which confront road users are ice, snow, wind and fog. These are exacerbated whenever they occur in combination and in darkness. Ice and snow can cause as much disruption as fog, but they can be predicted with a fair degree of certainty, they can be monitored *in situ*, and can be cleared or ameliorated by gritting and by the use of snow-ploughs. Wind conditions are usually forecast accurately, and the hazard can be reduced locally by the careful design of road sections and bridge structures, and by the use of wind-breaks. Fog is the least predictable of these hazards, both in terms of its variability over time and from place to place. It is expensive to monitor visibility in real time beside a motorway. There is no known means of preventing the occurrence of fog, or of dispersing it at an acceptable cost once it has formed. Fog can persist in some localities for many hours at a time.

Accurate knowledge of local fog hazards along proposed new roads and motorways is necessary to ensure the safety of eventual road users, and to ensure that the road is economically viable. Thick fog tends to reduce traffic volumes and to increase the likelihood of accidents to those vehicles on the road. If several alternative routes for a new road scheme are being considered by planners, those routes which will suffer from a high incidence of thick fog will tend to carry less traffic, suffer higher accident rates and will experience higher maintenance and construction costs than the others, and thus from a planning point of view will be less economic.

It should be borne in mind that the local variability of fog (and climatological factors in general) is but one of a large number of inputs to the overall decision-making process for new road and motorway schemes. In the past it has all too often been assumed that climatological differences were not significant; this is not necessarily so. The author has provided advice on fog (and other meteorological hazards) for the Department of Transport, the Welsh Office and other organizations for nine new road and motorway schemes; these have included the M20/A20 Dover–Folkestone (Channel Tunnel) Link, the A55 North Wales Coast Road, the Stoke–Derby M1/M6 Link Road, and part of the M25 London Outer Orbital motorway. Such advice has been used for public consultation exercises and public enquiries by the planners.

6.1 FOG TYPES AND FOG CLIMATOLOGY

6.1.1 Visibility

Clearly the principal effect of the presence of fog is to reduce visibility. Visibility is officially defined as: 'The greatest distance at which a black object of suitable dimensions may be seen and recognized against the horizon, sky, or in the case of night observations, could be seen and recognized if the general illumination were raised to the normal daylight level' (Meteorological Office 1969). Visibility is the term employed when observers determine the distance to black objects, visibility boards, landscape features on the horizon or lights (at night). When optical instruments are used, the term meteorological visual range or optical range is more commonly used. For all practical purposes these are the same; slight differences arise from the assumptions concerning light scattering inherent in the operation of the instrument and in the variability of visual acuity from person to person (Jiusto 1981).

Daylight visibility depends upon several factors, the most important of which are:

(a) the transparency of the atmosphere;
(b) the degree of contrast between an object and the background against which it is observed;
(c) the position of the sun; and
(d) physiological factors involving the observer, such as defective or exceptionally acute vision.

The aim in routine meteorological observations of visibility is to eliminate as far as possible all factors other than the atmospheric transparency. This is governed by the constituents of the air, which include not only water droplets but also dust, smoke and drifting snow kept in suspension by turbulence. Smoke and pollution may both prove important factors on occasions in reducing visibility on roads. It must be said however that 'Good quality visibility data are scarce, compared with data for, say, temperature and rainfall' (Lawrence 1976).

Fog is defined as a state of atmospheric obscurity in which visibility is *less than 1 km*, irrespective of whether the obscurity is produced by

Table 6.1 *International classification of visibility (Meteorological Office 1969)*

Visibility	*Description*
Less than 40 m	Dense fog
40–200 m	Thick fog
200–1000 m	Fog
1–2 km	Mist (if mainly due to water droplets)
	Haze (if mainly due to smoke or dust)
2–4 km	Poor visibility
4–10 km	Moderate visibility
10–40 km	Good visibility
over 40 km	Excellent visibility

water droplets or solid particles. If the visibility is *less than 200 m* the fog is described as *thick fog*, and if it falls *below 40 m* it is described as *dense fog* (Meteorological Office 1969). The international classification of the various categories of visibility is given in Table 6.1. Thick fog represents the most important visibility range for road users (Moore and Cooper 1972); work by the Transport and Road Research Laboratory has shown that drivers become particularly sensitive to the hazard when the visibility falls below 150 m (White and Jeffery 1980).

Fog forms when condensation occurs near ground level, and the major difference from cloud is that the base of the fog is found at or close to ground level. Horizontal visibility may vary considerably with the direction of view from any given point, depending on surface winds and local environmental factors. Visibilities in general are much lower for winds from the land than for winds off the sea, owing to greater concentrations of particulate pollution present in the atmosphere.

6.1.2 Types of fog

The first classification of fog types was devised by Willett (1928); this was extended by Byers (1959) and it is now considered that some fifteen types and subtypes of fog exist around the world. There are four basic types of fog which occur in the United Kingdom and Europe, as follows.

(a) Radiation fog

This occurs on cloudless nights when the earth's surface radiates long-wave radiation to space and loses heat, thus chilling the air adjacent to

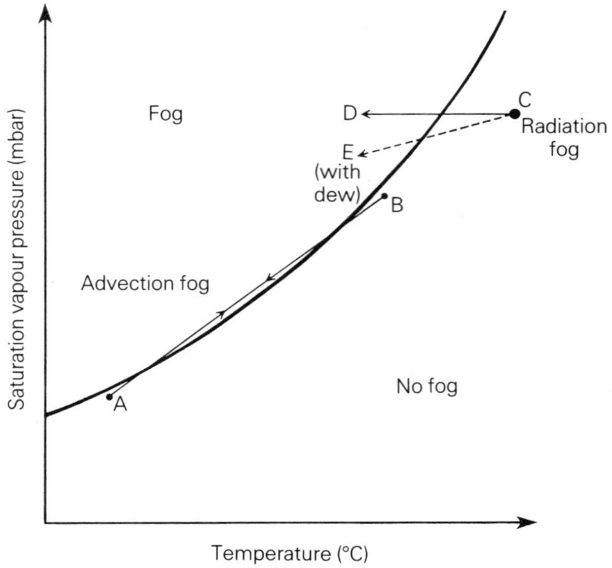

Figure 6.1 Fog formation processes (see text for explanation of letters).

the surface below its dew-point temperature (line C–D in Figure 6.1) and forming local fog. As the cooling occurs, a temperature inversion (temperatures colder near the ground than higher up) develops near the ground, and this together with the very light winds inhibits any downward transfer of heat by turbulence from warmer layers of the atmosphere aloft.

(b) Advection fog

This occurs when warm, moist air moves across a region with a cold surface (eg a snow-covered land surface or a cool sea). The air may be cooled below its dew-point temperature (line B–A in Figure 6.1) and a layer of fog may result; this is more important in coastal areas (where it is known as *sea fog*) than inland.

(c) Frontal fog

This occurs when a warm front with its associated drizzle advances over a cold underlying surface. When the drizzle falls through the cold unsaturated air, some of it evaporates, increasing the dew-point temperature of the unsaturated air and cooling the air: two processes leading to saturation of the air mass and reduction of visibility. A lowering of the cloud base and fog results.

(d) Upslope or hill fog

This occurs when air moves upwards over rising ground; the normal decrease of temperature with height may cool the air to its dew-point temperature and fog may form as a low-level 'cloud'.

The three main processes responsible for fog formation are therefore:

(a) cooling of air to its dew-point temperature (radiation fog and hill fog);
(b) mixing of moist air parcels of different temperatures (advection fog); and
(c) addition of water vapour to the air (frontal fog).

Clearly more than one of these processes can and does occur in the development of a given fog, but usually one mechanism is dominant. In terms of temperatures alone, warm fogs occur when the temperatures of the water droplets are above 0 °C; supercooled fogs occur where the droplets are in equilibrium with the atmosphere at a temperature which is equal to or less than 0 °C, and ice fogs occur when they are composed of ice crystals at temperatures considerably below 0 °C (these last are comparatively rare).

The seasons of occurrence, areas affected by the different types of fog, and their modes of formation and dispersal, are summarized in Table 6.2. All the types of fog are essentially governed by the prevailing synoptic situation. Sea fog is normally restricted to coastal areas, hill fog is a feature of the uplands, while advection fog and frontal fog tend to be widespread in occurrence; radiation fog, in contrast, tends to be localized and patchy in occurrence. Very little information exists on the climatology of the different fog types, for as Wheeler (1986) has stated: 'fogs generally have been studied less than many other aspects of our weather'. One of the few studies of fog climatology has been made of the central Ukraine, where Prokh (1966) suggests that advection fogs account for 59% of all foggy days, radiation fogs for 34% and frontal fogs for 7% of the overall total. The major types of fog affecting highway design and operation in the United Kingdom are radiation fog, and to a lesser extent sea fog and hill fog. These will now be discussed in detail, for it is important to understand their characteristics if the hazards which they produce are to be avoided.

6.1.3 Radiation fog

Radiation fog is the most serious and persistent type of fog hazard for the road user as it tends to be localized and dense, producing unexpectedly low visibilities which can cause trouble even to the most attentive driver. Radiation fog tends to form in damp, sheltered valley

Table 6.2 *Summary of fog types affecting the United Kingdom (after Codling 1971)*

Type of fog	*Season of occurrence*	*Areas affected*	*Mode of formation*	*Mode of dispersal*
Radiation fog	Autumn and winter	Inland areas, especially river valleys and low-lying damp ground	Cooling due to radiation from the ground on clear anticyclonic nights in conditions of light winds	Heating of the ground by the sun or increased wind
Advection fog				
(a) over land	Winter and spring	Often widespread inland	Warm air cooled by movement over cold land	Change in airflow or heating of the land
(b) over sea (sea fog)	Spring and early summer	Sea and coastal areas	Warm air cooled by movement over cool sea surface	Change in airflow or heating of the coast
Frontal fog	All seasons	Inland, especially high ground	Rain/drizzle from warm airmass falling into cold airmass close to saturation	Increase in intensity of the circulation or passage of the warm front
Hill fog	All seasons	High ground	Low cloud forming beneath the summit of hills	Change in circulation

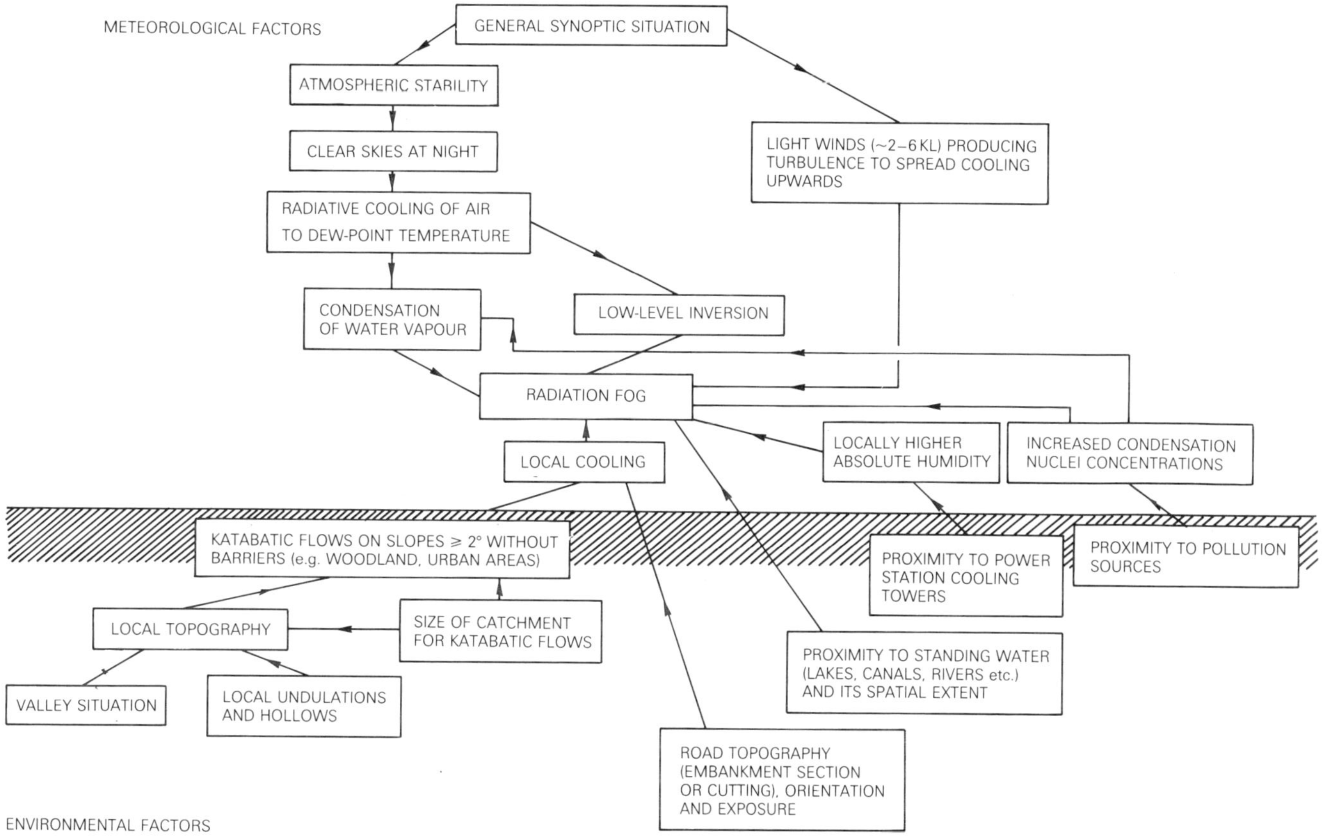

Figure 6.2 Meteorological and environmental factors influencing the local variability of radiation fog.

situations. The meteorological conditions favourable for its development (see Figure 6.2) may be summarized as: a clear night sky to allow maximum loss of radiation from the ground, preferably during long winter nights; moist air at sunset (most often found in autumn and winter), especially after rain or near open water; and a light wind (approximately 1–3 m/s or 2–6 knots) to give sufficient turbulence to spread the cooling upwards – anticyclonic conditions are favourable for this. Recent research (Findlater 1985) has drawn attention to the importance of a decrease in low-level turbulence at the onset of radiation fog formation, with the wind decreasing from 1–2 m/s to 0.5 m/s or less.

For radiation fog to form, the air adjacent to the ground must be cooled. There are two ways in which this can happen. The first is by turbulence, so that the air cooled by conduction with the ground mixes with warmer air aloft. The depth through which this takes place increases with the wind speed; it is usually up to 50–100 m deep. The second is by radiation from one layer to another or from the water droplets in suspension; this is a complex process, but the final result is a cooling effect up to about 100 m.

On a night during which radiation fog subsequently forms, observations have shown that not only does the temperature at 2 m above the ground decrease, but so does the dew-point. This is to be expected since dew is being deposited on the ground, thus removing water vapour from the air (line C–E in Figure 6.1). If the air is to become saturated, the temperature must fall more quickly than the dew-point, but observations show that this is still a slow process; it may take several hours for the relative humidity to rise from 95 to 100%. The

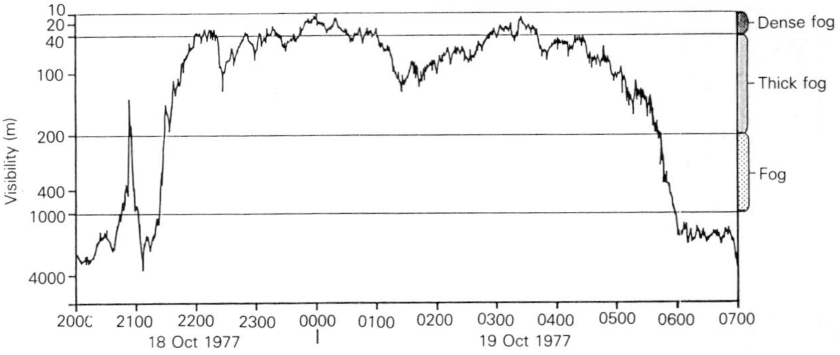

Figure 6.3 Part of a chart from a transmissometer showing an episode of radiation fog recorded at Wetherby, Yorkshire, October 1977. Note the speed with which the dense fog developed between 2100 and 2200 h.

temperature at which fog eventually forms, the fog point, may be several degrees Celsius below the original dew point of the air. Fog formation is usually rapid once started in air that is free from smoke pollution, and the visibility can fall from 3 km to 200 m or less in perhaps 10 min (see Figure 6.3).

Radiation fogs vary greatly in depth, but most are in the range 15–100 m (say 50–300 ft), but on extreme occasions the top may be at 200 or 300 m (600–1000 ft), while in relatively calm conditions the top may be only a few metres above the ground. Results from Cardington using acoustic sounders suggest that radiation fogs in Bedfordshire have typical depths of 60–240 m (Caughey *et al.* 1978). Table 6.3 suggests that the typical horizontal visibility within radiation fogs is some 100 m, although this average disguises enormous variability over space and time.

The highest frequency of radiation fogs occurs about an hour after sunrise. The slight increase in turbulence due to the early morning sun may cause existing fog to thicken or a sudden formation to occur when only dew had formed before. The theory may not be completely free from objection, but nevertheless the fact is important, and consideration should be given to the sudden formation of fog just after sunrise following a calm, clear night. The fog formed in this way in the morning or overnight then requires further surface heating or increased turbulence before it will dissipate (Jiusto and Lala 1980). In winter when insolation values are low, the clearance may be long delayed and a thick fog is likely to persist by shutting out the heating of the sun; in summer, radiation fog is infrequent and unlikely to persist much after sunrise. Because of the long nights and generally low land temperatures, the winter half of the year is most susceptible to radiation fogs.

Table 6.3 *Characteristics of radiation fog and advection fog (after Jiusto 1974)*

	Radiation fog	*Advection fog*
Average droplet diameter (μm)	10	20
Typical droplet range (μm)	5–35	7–65
Liquid water content (mg/m^3)	110	170
Droplet concentration (per cm^3)	200	40
Vertical fog depth (m)		
– typical	100	200
– severe	300	600
Horizontal visibility (m)	100	300
Spatial extent	Localized, patchy	Widespread

Radiation fog only forms when the precise combination of meteorological conditions shown in the upper half of Figure 6.2 is present. Its patchiness and local variability are also influenced by local environmental conditions which can accentuate the fog in certain localities rather than in others, producing the so-called fog 'black spots'. These factors are shown schematically in the lower half of Figure 6.2.

Local topography is important as it has the potential to generate flows of cold air, called katabatic flows. These may account for the fog often being erratic in development and localized in extent. As air in contact with the surface cools, it increases in density and may move slowly downhill, seeping into hollows and valley bottoms, called recipient areas (Hogg 1965). In these areas it may stagnate *in situ*, lose heat by radiation to the clear sky above, cool to below its dew-point temperature and produce a pocket of local fog. The land surface must have a gradient of over 2° for such a flow to develop, with no large-scale barriers to air movement, such as dense woodland or urban structures. The flow of cold air is normally shallow, confined to some 5–10 m or so above ground level; the gentle flow, although slow (perhaps 2 m/s) may have a large aggregate effect by the end of the night. Flows resulting from a single night's cooling in this country have depths within the range 30–150 m (100–500 ft), but some may be much shallower and others considerably deeper. This means that although raised stretches of road and motorway across major river valleys may occasionally remain clear of fog, they may be covered at times, depending upon the height of the road above the surrounding terrain and the depth of the fog. The size and geometry of the catchment from which such cold air drainage can accumulate is important in determining the depth of the eventual fog pocket.

The presence of local *standing water* (lakes, reservoirs, canals, rivers, saturated water-meadows, flooded gravel workings etc.) may also increase the susceptibility of a location to radiation fog. Because of the high specific heat of water, the area adjacent to the water body will tend to have lower maximum temperatures and higher minimum temperatures than the surroundings. This means that in conditions of light wind or calm, the air adjacent to the water will be slightly cooled on average in summer and slightly warmed in winter. Evaporation from the water bodies will raise the local absolute humidity of the air and will produce earlier saturation and condensation into fog. Radiation fogs will therefore develop earlier and linger in these areas.

The presence of local *power-station cooling towers* may raise the local absolute humidity of the air. Little pollution is emitted at the local level (the warm plumes from the tall stacks are emitted at high levels and disperse over a wide geographical region downwind), but the cooling towers emit sizeable quantities of water vapour into the atmosphere. The mist from the cooling towers tends to be worst in cold weather (the time of maximum frequency of radiation fogs) for three reasons: (a) the airflow in the towers tends to increase when the incoming air is cold; (b) the demand for electricity is high, hence the stations are working to capacity; and (c) the evaporating power of the surrounding air is reduced in cold weather so that local mists and fog may result and persist (Meetham 1964). Work by Vogel and Huff (1975) in the mid-West of the United States has shown how power-station cooling ponds can both initiate and enhance local fogs, especially in winter; such cooling ponds can often be seen steaming in cold weather.

The presence of local *woodland* often acts as a barrier to katabatic flows and thus influences cold-air drainage patterns, but it has two other important influences on radiation fog development. Dense woodland tends to have higher humidities than open country (particularly after rain or drizzle has occurred, where the leaves have intercepted water droplets which subsequently drip through the canopy), but temperatures tend to be ameliorated (lower than outside in summer and warmer than outside the woodland in winter). With lower wind speeds, more shading from the sun and higher humidities, mist and fog tend to linger in woodland, once formed. Its importance is a function of the size, height and density of the woodland.

Finally, *urban areas* exert a twofold influence over radiation fog formation in their environs. Urban structures tend to act as obstacles to any katabatic flows and thus retard and distort their development. The development of the nocturnal heat island under clear skies at night in winter also inhibits cold-air drainage. Lee (1987) has shown how, under such conditions, fog-free urban clear islands can develop within the fogs of the Central Valley of California. Urban areas are also important sources of atmospheric pollutants. Smoke and other aerosol particles resulting from the combustion process may reduce visibility levels within fogs by thickening and contaminating them. The results of several studies (e.g. Martin 1974; Unsworth 1979) suggest that the threshold concentration at which smoke particles begin to affect visibility is in the region 100–200 mg/m^3. However, at lower concentrations, they provide increased numbers of condensation nuclei and therefore increase the concentrations of water droplets within the fog. In general it has been shown that fog is more common in the suburbs of major urban areas than in the central areas (Chandler 1965).

The meteorological and environmental factors responsible for radiation fog formation are thus complex, and interact as shown schematically in Figure 6.2. Because of its patchiness and speed of formation (see Figure 6.3), precise forecasting of radiation fog and the resulting visibility levels over time and space is extremely difficult. Jack (1966) suggested that the 'percentage usefulness' of 12-hour forecasts of visibility levels of 200–500 m may be as low as 50–65%. This is a pity, for radiation fog is particularly important in producing the so-called 'fog black spots' on motorways and major roads in this country. An unexpected patch of thick radiation fog was probably the cause of the major collision on the M25 in Surrey in December 1984, which involved 26 vehicles and resulted in nine deaths. At the resulting inquest the coroner said that the crash was caused by 'an exceptionally thick and totally unexpected patch of fog'. Local environmental influences might well have been the catalyst for its local formation.

6.1.4 Advection fog

Advection fog occurs when warm, moist air moves over a cold land or sea surface and is chilled to its dew-point temperature to produce a layer of fog. Over land areas, advection fogs tend to be winter phenomena, developing when air has moved from over a warm sea surface to a relatively cool land surface (particularly if snow-covered). They tend to be regionally widespread and relatively uniform, in contrast to radiation fogs.

Sea fog is a form of advection fog which develops when moist air flows from a warm land or sea surface across a sea surface which is colder: one whose temperature is below the dew-point of the air. The wind which necessarily accompanies advection fog is the agent for spreading the cool lowest layers of air through a greater depth. The stronger the wind the greater the depth of the layer in which cooling takes place, and hence the greater the rate of cooling necessary to produce a fog. If the wind is too strong or the rate of cooling too small, or both, only low cloud (stratus) is formed, and there may even be no condensation at all.

Sea fog is essentially a spring and early summer phenomenon, occurring mostly between April and August when the air/sea temperature contrasts are greatest. It develops with moist south-westerly winds flowing across a progressively cooler sea surface as Britain is approached. It tends to affect the English Channel and Irish Sea coasts more than the east coast. Fog drifting inland from the North Sea over the coasts of north-east England and eastern Scotland in the spring and early summer is referred to locally as 'haar'.

It is the air/sea temperature difference which determines both the stability of the overlying air and the likelihood of fog formation. In summer, the air above the sea surface is cooled from below and becomes more stable as it moves over cooler waters. This reduces the tendency for vertical mixing and the resultant loss of moisture upwards. In winter the sea tends to be warmer than the air which enhances instability in the lower layers, and hence vertical mixing and loss of moisture upwards occur. The differing summer/winter stability factor is supplemented by differing wind-speed regimes in the two seasons. Winds in general are lighter in summer, again emphasizing summer as the most important season for sea-fog formation. Some wind is necessary to maintain advection but it must not be too strong. The optimum windspeed is 3–5 m/s (7–10 knots). If the wind is much greater than this, the fog may either dissipate (due to the vigorous stirring of the atmosphere that results) or rise to form a layer of low stratus with a low cloud base.

The thickness of a bank of sea fog is determined by the strength of the wind and the atmospheric stability. If the surface cooling necessary to produce the fog initially is sufficiently rapid to outweigh the turbulent mixing of fog-laden air with unsaturated air above, then the layer of fog may be up to 100–200 m thick.

The forecasting of the onset and dispersal of sea fog is more difficult than the forecasting of radiation fog or advection fog over land because, with the latter, more accurate estimates can usually be made of the surface temperatures and other parameters involved. Furthermore, the diurnal range of sea temperature is very small (especially in the open ocean away from the coast) and the temperature of the air overlying the sea only changes markedly if it is moving from a warmer to a colder sea area or vice versa. Formation of sea fog depends on only small changes in air temperature, upon the magnitude and sign of the temperature difference between the sea and air, as well as upon wind speed. It may be slow to form and slow to clear.

Over the sea, sea fog has little or no diurnal variation as a consequence. It may persist for hours or even days. However as it spreads inland it is usually dispersed by sunshine and the higher land temperatures, especially in the summer. It may return at night by spreading in from the sea, its formation often being assisted by the radiational cooling of the land. It may be further complicated, however, by interactions with local land/sea breeze circulations (Findlater 1985). On windward coasts it can persist all day with cool, clammy conditions, whereas perhaps 10 km inland the day may be bright, warm and dry: an important factor when planning the location of a new road in a coastal location susceptible to sea fog in the summer months.

There is little published material on the inland penetration of sea fog. Pratt (1968) concluded from a study of the haar on the north-east coast of England that the maximum distance reached inland by sea fog is some 10 km. The extent of landward penetration of sea fog is highly variable and depends on local wind and temperature conditions, together with the topography and geometry of the coastline; a maximum distance of 6–10 km inland would apply in reasonably flat terrain, and the distance is normally less.

6.1.5 Hill fog

Hill fog is simply low cloud, which to an observer above the cloud base appears as fog. It arises from the air's cooling as it is forced upwards over ranges of hills or mountains, and on cooling to its dew-point temperature condensation and cloud results. Hill fog is clearly an important hazard in upland areas traversed by major roads, and visibility within the fog tends to be variable.

Pedgley (1967) has suggested that when winds are strong, so that the lower layers of the atmosphere are well mixed by turbulence, then the height of the cloud base is determined by the humidity of the airstream measured near the ground, and can be expressed by the empirical relationship:

$$\text{Cloudbase height in km} = \text{dewpoint depression (°C)}/8$$

In a southerly airstream ahead of an approaching depression, the surface dew-point depression becomes quite small in regions near the coast, commonly less than 2 °C, with the result that a sheet of cloud may develop with its base only about 250 m above sea level.

Hill fogs may be exacerbated by the presence of rain and darkness. The resultant hazard is a function of the altitude of a given location, its exposure, distance from the sea and the characteristics of the prevailing airmass. As the airflow may be affected to a great depth, hill fog may be deep and it may persist for lengthy periods of time.

6.2 FOG, TRAFFIC VOLUMES AND ROAD ACCIDENTS

The presence of fog, especially thick fog, influences road traffic in two ways: it reduces the volume of traffic on the road, and it increases the risk of accident to those road users present. Time is normally lost through delayed traffic and generally reduced speeds in fog, while driving itself is more stressful when fog is known to be present. Fog is that aspect of weather on major roads and motorways which drivers fear most. It is an especially serious hazard to vehicles when visibility varies over short distances or periods of time at so-called fog 'black spots' or fog-walls. A report by the OECD (1986) concludes that 'driving under conditions of reduced visibility, especially during the hours of darkness and twilight, is a demanding and relatively dangerous task'.

Thick fog has the effect of reducing traffic volumes on motorways by some 20% compared with mean conditions, although this figure is variable and there have been few studies relating to the problem. Tanner (1952), investigating traffic flows on the A4 near Heathrow Airport, suggested that volumes were 20–40% below normal on weekdays and 40–60% below normal at weekends in foggy weather. Codling (1971) analysed traffic flows in 1964, and concluded that in thick fog they were reduced by some 19.6% on motorways, 21.1% on Class A roads, and 22.2% on Class B roads, but these means hide considerable variations on a daily basis. The reduction will vary according to the day of the week, the time of day and the state of the road (whether snow and ice are also present). When thick fog is present or is likely to occur, journeys may be postponed, travellers may resort to other forms of transport, or the journeys may be re-routed. However to quote from Moore and Cooper (1972): 'With present knowledge, one cannot make a statement in the form "at a trafffic flow level of *x* vehicles per lane per hour the accident rate per kilometre travelled is *y* per cent greater (or less) in fog than when there is no fog"'.

In a study of the effect of thick fog on traffic flows on Californian freeways, Kocmond and Perchonok (1970) demonstrated how the hazard both slows down traffic and reduces the number of vehicles on the road. They showed that speeds were reduced by between 5 and 8 mi/h (8 and 13 km/h) in thick fog, and from other data collected on six different occasions on a rural highway near Elmira, New York, they showed speeds to be reduced by 4–5 mi/h (6.5–8 km/h) in thick fog; this is not a large reduction.

The most serious aspect of fog on motorways is clearly the problem of the increased risk of serious accident for the road user. The motorway network (which is generally densest in that part of the country which suffers the highest incidence of thick fogs) has a generally good and improving safety record, but accident rates in conditions of thick fog tend to increase, with a higher percentage than normal involving multiple collisions. On average, from 1970 to 1974, only 1.4% of all serious-injury accidents on the roads of the United Kingdom occurred in fog or mist (such conditions are relatively rare, hence the low percentage); on motorways however, 6% of accidents occurred in thick fog (Codling 1974) – a very significant difference which can perhaps be attributed to the higher driving speeds, the lack of road intersections and the limited scope for manoeuvring imposed by the type of road system. From 1976 to 1985, 0.51 to 1.44% of all serious injury accidents each year occurred in fog (TRRL 1987). In general, motorway accidents in fog are likely to be more serious and are more likely to be multiple collisions, with more casualties per accident than those occurring in other weather conditions (Johnson 1973). During 1985 for example, 47% of accidents in fog on motorways and A(M) roads involved more than two vehicles, compared with 19% in clear weather; on all other roads these percentages were 12% and 7% respectively (TRRL 1987).

In 1974, fog was estimated to have cost over £12 million on the roads of Great Britain, and a single multiple collision cost £$\frac{1}{2}$ million at these prices, respectively equivalent to £60 million and £3 million at 1990 prices. These figures include the cost of medical treatment, damage to vehicles and property, and administrative costs of police, services and insurance, but do not include the cost of delays to vehicles not directly involved in the accident; such delays to ambulances and fire engines travelling to and from the scene of the multiple accidents on motorways are themselves an important aspect of the hazard. Two examples of this type of multiple pile-up illustrate the magnitude of the problem. On 13 September 1971, 200 vehicles were involved in an accident on the M6 at Thelwall; 10 people were killed and 70 injured. On 13 March 1975, an accident between Junctions 17 and 18 on the M1 involved 204 vehicles over some 23 km of carriageway, and led to 6 deaths and 28 serious injuries. Many similar examples could be quoted from Europe and North America.

Such multiple collisions are usually extensively reported in the press, often beneath headlines referring to 'motorway madness'. They are usually caused by the sudden braking of vehicles entering an unexpected patch of fog with the following vehicles running into the back of them in the reduced visibility. There are many theories concerning why these multiple collisions occur, but among the factors considered to be important are the following.

(a) The motorway network is at its densest in that part of the country most susceptible to thick fog (i.e. central England); see Figure 6.4.
(b) In general, speeds are greater on motorways than on other roads because of the lack of junctions and roundabouts.
(c) Having entered a motorway, a driver is committed to going on if he encounters fog; there is no possibility of stopping, turning back or turning down a side road.
(d) There is a general fear of losing the guidance of the rear lights or silhouette of a vehicle in front in conditions of fog. This reluctance of drivers to be a pathfinder or leader of a queue of vehicles promotes the formation of groups in which a high proportion of drivers follow closer than they need to maintain visual contact with the vehicle in front (Jeffery and White 1981).

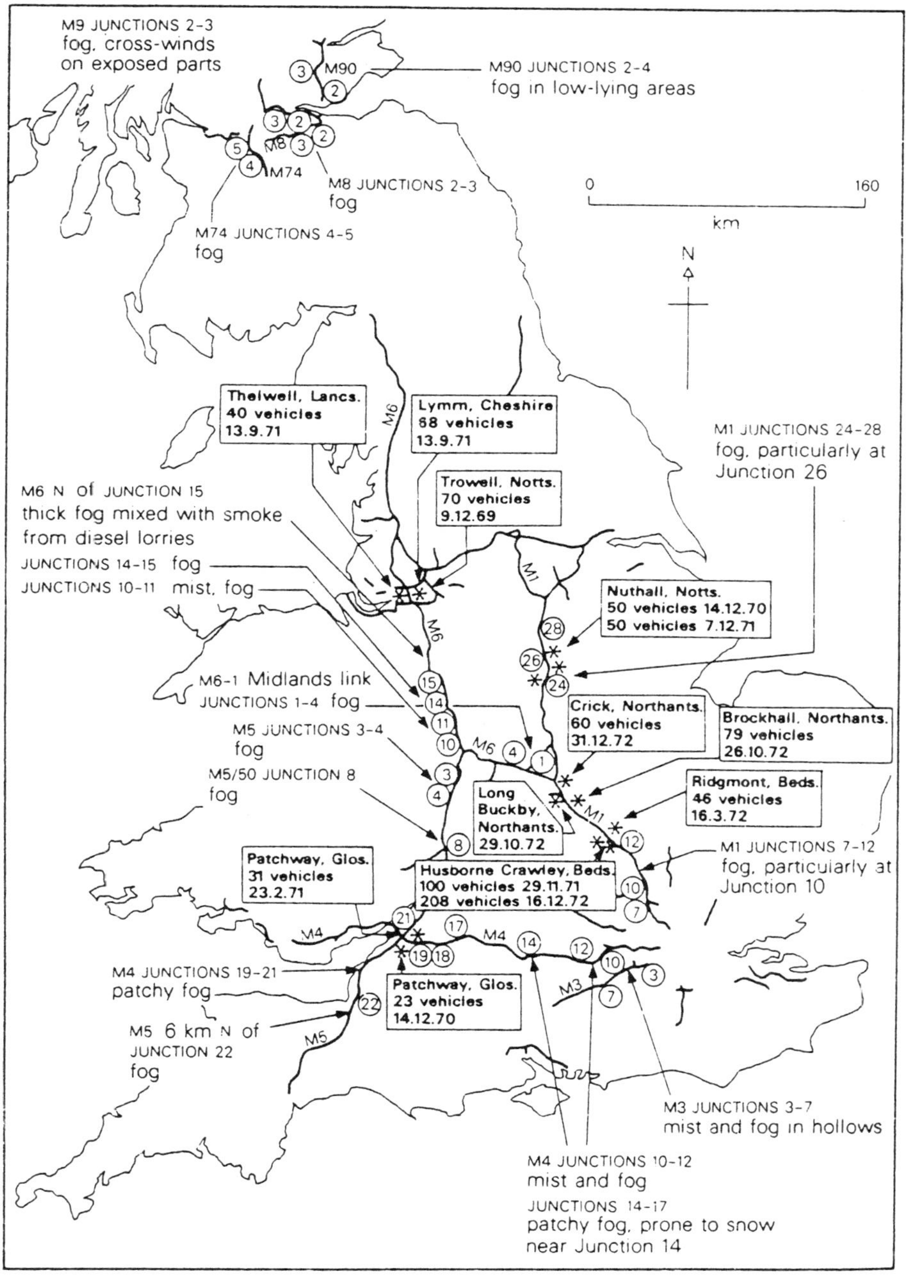

Figure 6.4 Fog-prone areas of the British motorway network and locations of multiple collisions (from Perry 1981).

(e) In fog, the driver may have to judge following distances and speeds without the normal visual clues, leading to disorientation. The fog tends to hide roadside features or reduce their visual contrast, thereby hiding their visual impact. It has been shown (Brown 1970) that objects appear to be further away than they really are in fog; hence the driver thinks he can see further ahead than is possible and underestimates his speed. The featureless environment reduces the impression of speed, and this 'speed hypnosis' is exacerbated by the fact that the driver cannot check his speedometer while concentrating on the road ahead. This theory is borne out by a detailed survey of 1780 vehicles driving in fog on the M4 in Berkshire during the winter of 1975/76 (Sumner *et al.* 1977). The measurements showed that *drivers did not reduce speed a great deal until visibility dropped to 100–150 m*, probably because they considered that the visibility enabled them to see far enough ahead to stop safely if required. However, the safety margin decreases as visibility gets worse. *At 50 m visibility, more than half of the drivers in lanes 2 and 3 were exceeding the speed at which they could stop within this visibility distance*, as defined by the Highway Code (see Figure 6.5). It seems that drivers are aware of the need to slow down in these conditions, but they do not slow down enough.

(f) The risk of a collision from the rear (particularly from a larger vehicle) may keep drivers moving faster than if they had been on ordinary roads. This view was expressed in a letter to *The Times* of 28 March 1974:

'Dear Sir,

I try to avoid motorways in fog, but if caught out my reason for driving too fast is to keep myself well ahead of the forty-ton lorry behind.

Yours etc.'

(g) All sound is muffled in fog; hence the driver does not hear the sound of the collisions ahead.

(h) Multiple accidents on motorways may be promoted by a leading driver slowing down on entering a particularly dense patch of fog; following drivers may (unintentionally at first) react by accelerating to keep the intensity of the preceding car's silhouette and rear lights constant, in a situation where they should be slowing down. The drivers may perceive that they can control the possibility of collision in front with their own skill and reaction in applying the brakes, but seek to minimize the possibility of a rear-end collision (often from larger vehicles) by maintaining speeds which exceed the safety margin.

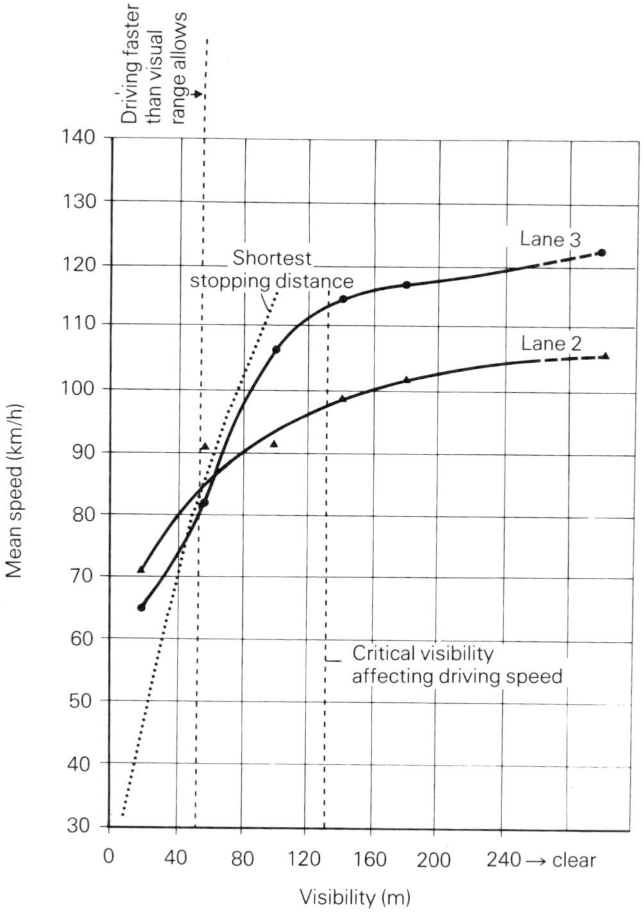

Figure 6.5 Mean speeds of private cars and light commercial vehicles.

6.3 THE ASSESSMENT OF LOCAL FOG CLIMATOLOGY FOR NEW MOTORWAY SCHEMES

Thick radiation fog is difficult to forecast with any accuracy; it is difficult to remove once formed; and it is not always feasible or possible to take remedial measures. The planners of new major road or motorway schemes are therefore increasingly concerned with the identification of locations which are likely to suffer a high incidence of thick fog (commonly called 'fog black spots'), with the hope of avoiding them.

The analysis of the local fog climatology for such schemes is not easy, particularly in complex terrain. A number of techniques have been employed, frequently used together.

6.3.1 Analysis of data from standard meteorological stations

The obvious source of visibility data is from the records of existing standard meteorological observing stations and climatological stations. The data are useful, particularly when obtained from sites maintained by trained personnel. Visibility is not easy to measure, however, especially at night. There are a number of problems in using such data as follows.

(a) Many of the most reliable meteorological stations which do report visibility regularly and accurately are airfields; by definition the sites for the airfields are normally flat, open sites, chosen to *avoid* foggy locations; hence the visibility data from these stations (often run by the Meteorological Office) may not be typical or truly representative of the worst of conditions in the surrounding area, particularly where this consists of undulating topography, or lies next to the sea or industrial complexes.
(b) The meteorological stations may be some distance from the area being studied.
(c) It is often impossible to interpolate the incidence of fog between observing sites, because of the known variability of fog (as a consequence there are very few maps of the incidence of fog published in the literature).
(d) The incidence of fog and thick fog has tended to decrease markedly during the last thirty years due to the impact of the Clean Air Act legislation, commencing in 1956, and the shift to the use of smokeless fuels both in the domestic and industrial sectors.

Figure 6.6 shows how the number of days per year with fog has changed over the period 1950–1983 for a range of 28 stations across Great Britain. The detailed data from which the map was constructed are included in Table 6.4. Most of the stations show a clear decrease in the incidence of fog, the exceptions being Wick and Tiree in Scotland, and Aberporth in Wales (all fairly remote coastal sites which experienced relatively little air pollution in the 1950s). Overall, if data for the decade 1974–1983 is compared with the data for 1950–1959, there has been an average decrease of 45% for the 28 stations. The most dramatic percentage reductions have occurred at Abbotsinch (Glasgow Airport), Tynemouth, Watnall, Elmdon, Finningley (north-east of Sheffield), Heathrow, Ringway (Manchester Airport) and Filton (Bristol): all sites

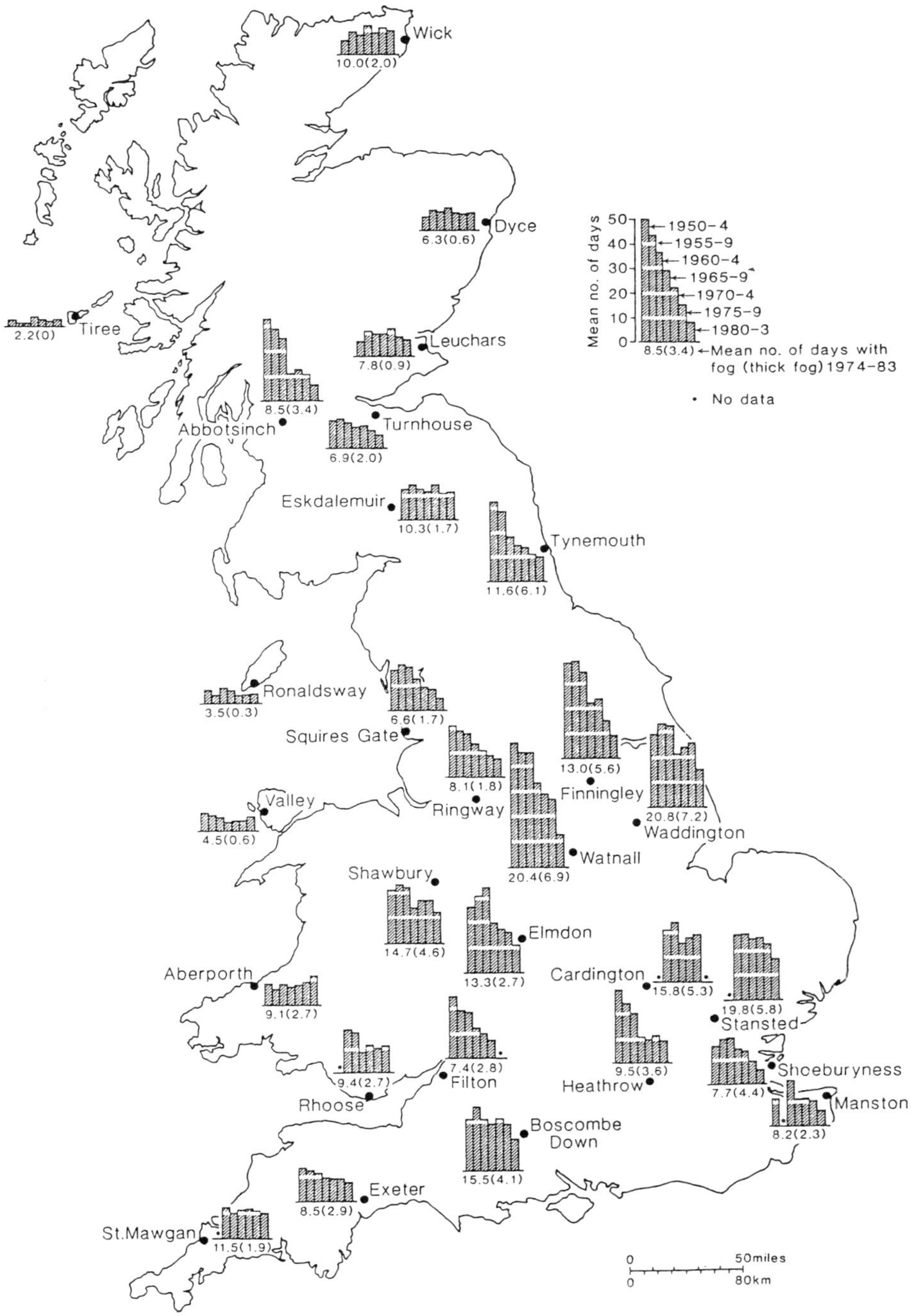

Figure 6.6 The spatial variation of fog over Great Britain, 1950–1983.

Table 6.4 *The changing incidence of fog in Great Britain, 1950–1983*

Station	*Average number of days per year with fog at 0900 GMT*							*% change 1974–83 compared to 1950–59*	*Mean days per year 1974–83*	
	1950–54	*1955–59*	*1960–64*	*1965–69*	*1970–74*	*1975–79*	*1980–83*		*Fog*	*Thick fog*
Wick	5.4	9.8	7.6	11.4	8.8	10.4	9.7	+31.6	10.0	2.0
Dyce	5.2	8.0	7.2	9.0	6.8	6.2	6.7	−4.5	6.3	0.6
Leuchars	6.4	10.4	9.2	9.6	11.6	8.2	7.0	−7.1	7.8	0.9
Turnhouse	11.6	12.0	10.8	8.8	9.4	7.2	5.7	−41.5	6.9	2.0
Tiree	2.6	1.4	1.4	3.8	2.6	2.0	2.7	+10.0	2.2	0.0
Abbotsinch	33.6	29.0	25.2	11.0	12.6	10.4	6.2	−72.8	8.5	3.4
Eskdalemuir	12.8	14.2	12.8	11.8	14.2	10.4	11.2	−23.7	10.3	1.7
Ronaldsway	5.6	3.4	6.2	5.2	3.2	3.4	3.7	−22.2	3.5	0.3
Tynemouth	32.6	28.4	18.2	14.8	14.0	10.6	10.0	−62.0	11.6	6.1
Waddington	29.0	33.5	33.0	21.5	24.0	25.6	15.5	−33.4	20.8	7.2
Shoeburyness	15.7	18.4	18.8	14.4	13.4	9.4	6.0	−54.8	7.7	4.4
Cardington	–	20.8	23.8	15.2	17.6	18.8	–	−24.0	15.8	5.3
Stansted	–	26.2	26.6	24.2	25.0	22.6	16.7	−24.4	19.8	5.8
Shawbury	21.8	23.8	22.4	14.0	17.6	17.6	12.5	−35.5	14.7	4.6
Watnall	49.2	45.8	45.8	33.2	29.0	27.0	13.0	−57.1	20.4	6.9
Elmdon	26.8	30.2	34.2	20.0	17.4	16.2	10.5	−53.3	13.3	2.7
Finningley	38.7	39.5	35.0	22.8	24.2	15.2	9.5	−66.8	13.0	5.6
Heathrow	29.6	24.0	20.0	10.4	9.6	11.0	9.0	−65.3	9.5	3.6
Manston	11.0	–	18.5	10.6	10.8	10.0	6.2	−25.5	8.2	2.3
Boscombe Down	20.8	25.2	20.2	18.6	20.4	18.2	12.7	−32.6	15.5	4.1
Ringway	20.2	18.2	17.8	13.8	10.4	8.8	7.0	−57.8	8.1	1.8
Squires Gate	16.2	18.8	17.6	13.0	9.6	8.4	4.5	−62.3	6.6	1.7
Valley	7.2	6.0	5.2	3.2	4.0	4.0	5.5	−31.8	4.5	0.6
Filton	25.0	19.2	18.8	12.6	10.0	7.6	–	−66.5	7.4	2.8
Aberporth	8.6	5.6	8.2	7.2	7.8	8.0	11.5	+28.2	9.1	2.7
Exeter	13.8	12.4	11.4	9.6	9.4	9.4	8.7	−35.1	8.5	2.9
Rhoose	–	17.6	16.0	8.0	10.8	9.8	10.5	−46.6	9.4	2.7
St Mawgan	–	12.5	10.0	11.6	11.8	10.2	10.0	−8.0	11.5	1.9

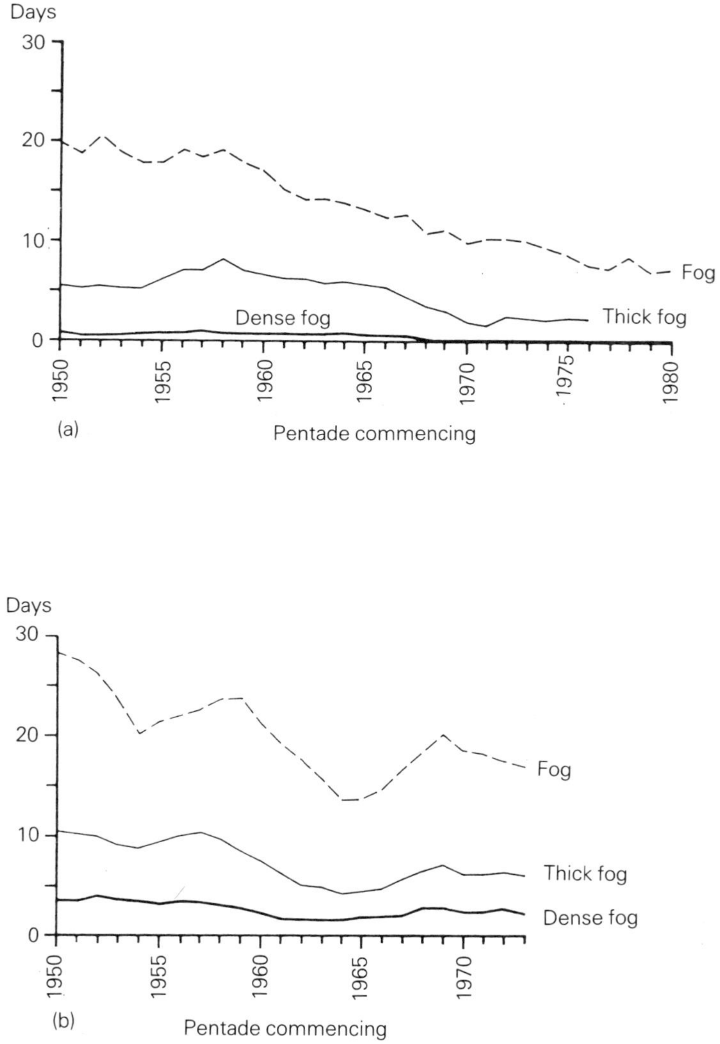

Figure 6.7 The changing incidence of fog, thick fog and dense fog at (a) Manchester Airport and (b) Speke, Liverpool, 1950–1984 (the data for Speke are for the period 1950–1977). The graphs represent five-year running means.

close to industrial/urban areas where Clean Air Act legislation has had a marked impact. Figure 6.7 shows how the incidence of fog, thick fog and dense fog has changed with time at Manchester Airport and Speke (Liverpool Airport). It is clear from Figures 6.6 and 6.7 that for much of the country the 'pea souper' fog is now a thing of the past, and that visibility records from the past cannot be used to indicate present or future visibility conditions.

6.3.2 Analysis of 'in situ' manual observations of visibility

The use of *in situ* information on the local fog climatology overcomes some of the difficulties outlined above. Manual visibility observations are normally obtained once daily (at 0900 h) at specific points of interest in the study area, using observations of visibility boards or of prominent objects in the landscape. Visibility boards are black boards, approximately one metre square, located with the sky as background where possible, with markers or pegs at standard distances from the boards to enable the observer to measure distance with accuracy. The sites can be chosen to monitor what are considered to be the worst zones along particular routes (although they must be reasonably accessible for the observers). Regular and systematic observations can be made (by paid helpers), together with comments and notes on whether the fog is patchy, localized, widespread, worsening or improving at the time of observation. The resultant data can provide a relatively low-cost *comparative* assessment of the local fog climatology (it is rarely more than comparative, because of the short period of time available for the survey).

A network of such observations has been used at 24 sites in the Stoke–Derby area as part of a fog survey in connection with a new motorway; at 12 sites in south Yorkshire as part of a fog survey along a corridor proposed for an extension of the M1; and at nine sites around the town of Newark-on-Trent in connection with a proposed bypass for the Midlands town. The problems with the observations (which can be minimized with care) are that their accuracy cannot be verified (there may well be differences in the quality of observations between individuals); there may be unfortunate gaps in the record on Sundays, holidays and times of illness; and the observations are normally made only once daily, which limits the amount of information available.

6.3.3 Analysis of data from automatic visibility recording equipment

Instrumental observations of visibility can be made using transmissometers and other instruments which provide a continuous chart record of visibility at a particular site (an example is shown in Figure 6.3). Equipment such as the Transport & Road Research Laboratory transmissometer measures the attenuation of a beam of light after it has passed through a known distance of the atmosphere, and this can be related to visibility. Other systems measure the degree of forward scatter or back scatter of light by fog droplets in the atmosphere (Douglas *et al.* 1978). The equipment normally provides continuous records of visibility at pre-determined locations of interest and the data are reasonably accurate and comparable. They are subject to the normal vicissitudes of automatic recording sensors, while blown snow and local insect life (attracted to the light) may occasionally interrupt the light beam and distort the data if care is not taken. The equipment is costly (thus only a small number of sites can normally be monitored) and it often requires major installations with protective fences if it is to be safe; these may give rise to local 'planning blight'.

The Department of Transport tested five different fog-detecting systems (two transmissometers, two sensors which measure the amount of back scatter of light by the fog droplets, and a sensor which measures the forward scatter of light) and installed equipment at thirty fog-prone sites along the M25 around London which became operational in 1990. The detectors automatically switch on matrix signals which display the word 'Fog' when visibility falls below 300 m. The system is fully integrated into the four existing M25 motorway control centres. It is hoped that after evaluation by TRRL similar systems will be installed on other motorways.

6.3.4 Visibility traverse data

A cheap solution to the problem of obtaining good comparative data at a number of locations over a compact area at times additional to 0900 hours is to use manual visibility observations from traverses through the fog made by a conscientious observer. The individual uses a fixed, predetermined route through the area (made in the reverse direction on alternate traverses to overcome the problem of visibility levels changing with time as the traverse proceeds), stopping at selected points *en route* to make detailed visibility observations. For this, prominent features in the landscape are used, with the information plotted on a 1 : 10 000 scale base map, while written comments are made at the time on the spatial and temporal variability of the fog in the locality. This allows a detailed, low-cost analysis of the spatial variability of visibility to be built up from a number of case studies of individual fog episodes.

The author has used data from 24 traverses through fog in the

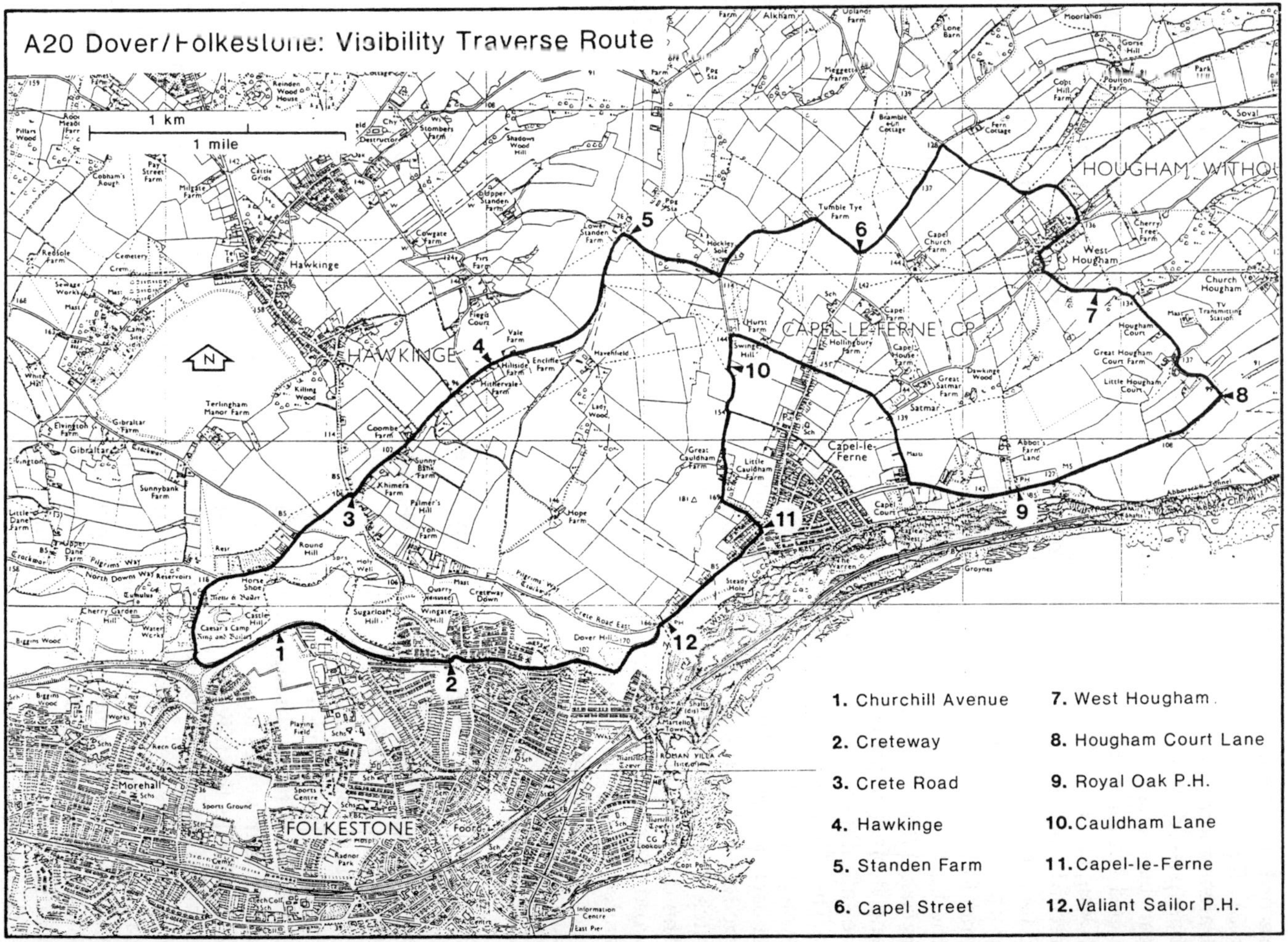

Figure 6.8 Visibility traverse route used to monitor the spatial variability of visibility for the M20/A20 Dover–Folkestone scheme, Kent.

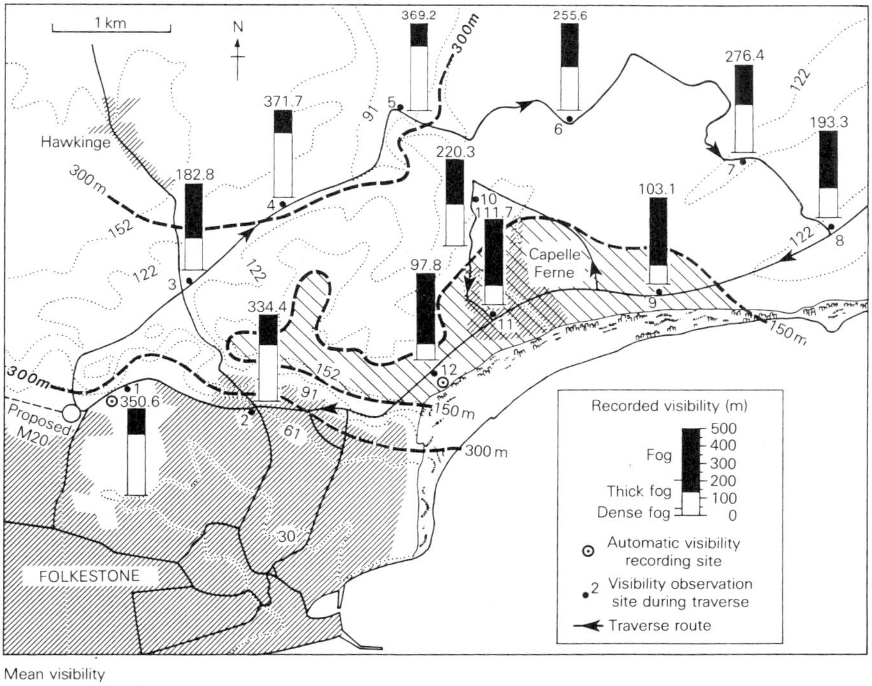

Figure 6.9 Mean visibilities recorded at each observation point during 24 visibility traverses in the Dover–Folkestone area. Note the figures of less than 150 m arond Capel-le-Ferne compared with the figures of over 300 m further north.

Dover–Folkestone area of Kent to evaluate the spatial extent of hill fog and sea fog in the locality (in the vicinity of the proposed Channel Tunnel terminal). Figure 6.8 shows the extent of the traverse, while Figures 6.9 and 6.10 indicate the mean visibility and minimum visibility recorded at each of the sites during the traverses. The lengths of the white bars used in Figures 6.9 and 6.10 represent the visibility levels, with site values given at the top of each column. The shaded area around Capel-le-Ferne emerges as a particularly fog-prone locality, with conditions much worse than those just two kilometres to the north. An independent check of the accuracy of the observations using data from two transmissometers at fixed points on the traverse route, revealed the accuracy of the manual observations to be of a high standard. The survey results allowed a clear evaluation to be made of the spatial variation of fog in the area.

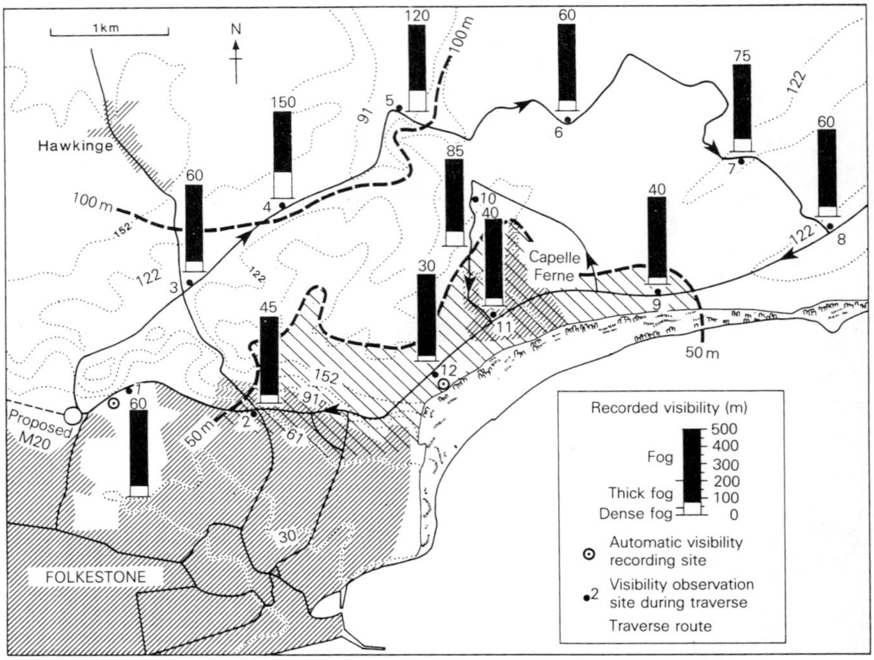

Figure 6.10 Minimum visibilities recorded at each observation point during the Dover–Folkestone visibility survey. Note the low values around Capel-le-Ferne and the higher values to the north.

6.3.5 Analysis of road accident data, police comments etc

If accident data are available from police records, it is sometimes possible to analyse the locations of those accidents which have occurred in fog in the area of interest. Patterns of clustering may suggest possible fog black spots. However, accident numbers are a function of traffic volumes along particular roads and not all potentially hazardous fogs will give rise to an accident. Not all accidents which occurred when the weather was foggy may have been caused by the fog. The analysis is thus useful in identifying possible problem locations along a stretch of road which can then be investigated further with a visibility survey, but the analysis cannot necessarily be used to compare different roads where the traffic volumes may differ considerably. This information may be reinforced by objective comments from reliable local sources who know the roads well, such as the police and the motoring organizations.

6.3.6 The Fog Potential Index Model

Fog, particularly radiation fog, is known to be very variable in occurrence and often localized in extent. It is known from experience that fog can be worse in some localities than in others because of local *environmental* influences. It is extremely difficult to produce accurate and useful climatic appraisals (or forecasts) of fog with a spatial resolution of less than, say, 10 km^2 from conventional data. Furthermore, it is not easy to suggest precisely where a fog detector should be located beside a particular motorway if it is to monitor the worst of the local fog. For these reasons it is particularly important to try and identify those local environmental factors which will accentuate the fog in certain localities, producing the localized fog black spots.

Such local environmental factors are indicated in Figure 6.2 and can be listed as follows:

(a) local topography: local valleys, hollows or dips which will pond up the coldest and densest air resulting from nocturnal katabatic flows;
(b) standing water and its spatial extent: lakes, reservoirs, rivers, flooded gravel workings, saturated water meadows and marshes;
(c) water vapour from anthropogenic sources; introduced into the atmosphere from local power station cooling towers and industrial complexes;
(d) particulate pollution sources: local factories and urban areas;
(e) woodland and forests.

The aim of the Fog Potential Index

The Fog Potential Index which the author has developed (with T.J. Chandler until 1977) is an attempt to quantify the effects of these local environmental factors considered important in producing local radiation fog with visibility less than 200 m. The general model can be used to identify those particular localities where the incidence of thick fog will tend to be greatest and where visibilities will tend to be lowest. The model thus aims *to assess the potential of a given location for thick radiation fog*. It is this thick fog which produces the greatest hazard for the road user, for not only does it tend to be erratic in development and localized in extent, but visibilities tend to be lower, on average, than in other types of fog. Radiation fog has a typical horizontal visibility level of 100 m, compared with the mean horizontal visibility within advection fog of 300 m (see Table 6.3).

The theory of the Fog Potential Index is based upon:

(a) published theoretical work on the importance of the variables affecting radiation fog;

(b) published field investigations of radiation fog (and katabatic drainage);
(c) previous case studies of fog on motorways;
(d) an investigation of the environmental factors contributing to accident blackspots in fog on the M6 in Lancashire and Cheshire, and the Midlands Links Motorways for the period 1970–1973; and
(e) personal work by the author on eight proposed new motorway and major road projects in different parts of England and Wales.

The model does not attempt to forecast fog, nor should the numerical indices derived from it be used to evaluate the complete fog climatology of a given location. Given the right meteorological conditions, hill fog is a function of altitude and exposure. Advection fogs tend to be widespread and regional in occurrence; thus they tend to occur at all the observing stations within an area when they do occur (although visibility levels may differ from station to station). Radiation fogs tend to be selective in their location, and are local rather than regional weather phenomena; the Index attempts to predict where these localities might be.

The Fog Potential Index $\mathbf{I_p}$

The Fog Potential Index I_p expresses the susceptibility to thick radiation fog of a given location p in terms of a numerical index I which has a range of values between 0 and 100. The indices are *comparative* (i.e. a value of 30.0 at point A, and one at B of 20.0, means that the ratio of the number of hours of thick radiation fogs at stations A and B will be in the ratio of 30 : 20, so that A will experience 50% more hours than B).

I_p is expressed in general terms by an expression of the form:

$$I_p = f\,(d_w,\ t_p,\ s_p,\ e_p)$$

Where

d_w expresses the distance of the location p from standing surface water and its spatial extent;

t_p is a function of the local topography at point p, incorporating such factors as the form of the local topography (hill or valley, slope or plain), the size of the likely catchment area for katabatic drainage towards the site, and whether the site is a recipient area for cold air flows or a donor;

s_p is a function of the road site topography at point p, expressing the form of the road at that point (whether cutting, bridge or embankment), and to a lesser extent the orientation of the road (influencing exposure);

e_p incorporates a general expression of any environmental features likely to help or hinder radiation fog formation in the locality, such as the proximity of pollution sources and the proximity of power station cooling or other 'artificial' moisture sources.

The variables which determine I_p are not of equal importance and weighting factors are applied to each. After further research, the form of the equation for predicting I_p at a point p is considered to be of the form:

$$I_p = 10d_w + 10t_p + 2s_p + 3e_p$$

Operation of the Fog Potential Index Model

Having described the governing equation for I_p and the importance of the various environmental inputs into the model, the expression then has to be applied as objectively as possible to the area of interest. Using maps at the scale of 1:25 000 and 1:50 000, together with detailed fieldwork along each route of interest, values of I_p are evaluated by assessing the appropriate value of d_w, t_p, s_p and e_p at each point. Values are normally mapped at either 0.5 km or 1 km intervals along suggested routes and at particular locations of interest (such as route intersections, river crossings etc.). This is done by awarding each variable a value between 0.0 and 4.0 (with half unit intervals or less where possible). The value is assigned as objectively as possible according to the importance of each factor at the point. A variable is only awarded a value above 0.0 if there is some genuine feature of the environment meriting it. Given that a maximum value of 4 can be assigned to each factor, and multiplying by the weighting factors already described in the formula, a maximum value of 100.0 can be assigned to any one location: i.e. $(10 \times 4) + (10 \times 4) + (2 \times 4) + (3 \times 4)$.

The numerical values so evaluated apply to the specific point locations and not necessarily to the area around them, for the importance of the various environmental inputs may change away from the point. It is not possible therefore to interpolate a linear increase or decrease between adjacent points, or to produce a useful contour map I_p without first evaluating I_p for many extra sampling points.

It must be made clear that the model is unpublished and still experimental. It was first developed to try to predict likely fog blackspots along proposed lines for the Stoke–Derby Link Road; when tested against a year of manual visibility observations during the winter of 1976/77 it produced 'reasonable agreement'. It was subsequently used

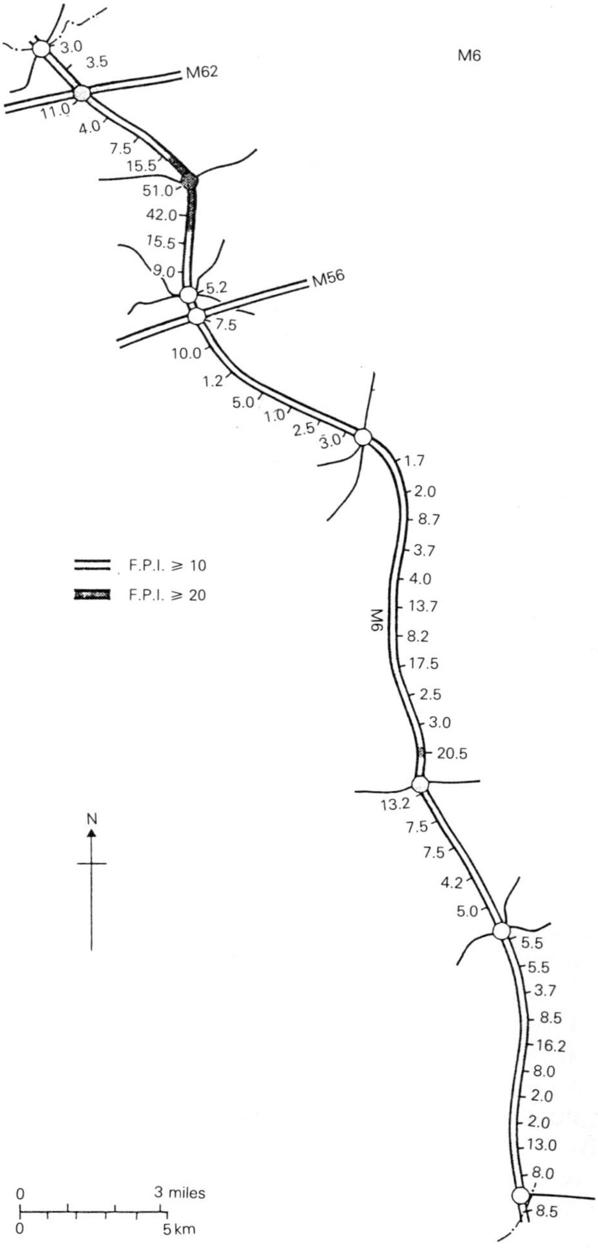

Figure 6.11 Spatial variation of the Fog Potential Index along the Cheshire section of the M6 motorway.

to predict the likelihood of radiation fog at two alternative locations for the crossing of the River Gade in Hertfordshire by the M25 London Outer Orbital Route, where it predicted the ratio of thick fog to be experienced at the two alternative valley crossings extremely accurately: it predicted the ratio of the incidence of thick fog at the two valley crossings to be 0.57 : 1, while subsequent transmissometer data at the two sites during 1978–80 showed a ratio of 0.52 : 1. In 1980 it was used to investigate and predict fog problems along selected corridors and along selected routes for a proposed extension of the M1 in Yorkshire between Kirkhamgate and Dishforth; the predictions agreed well with subsequent manual and instrumental observations of visibility. The technique has most recently been applied to the examination of fog black spots along all the major roads and motorways of Cheshire. Detailed maps of the Index have been produced and Figure 6.11 shows an example, portraying the Index along the Cheshire section of the M6; the highest values of I_p occur at the Thelwall viaduct where the motorway crosses the Manchester Ship Canal and the River Merey.

In 1985 the Meteorological Office produced a report for the Department of Transport on *The susceptibility of fog on the M25 motorway* (Met. Office 1985). The Fog Potential Index was used in the report to evaluate the problem of fog along the motorway with a view to installing fog detectors at the most fog-prone locations. The Index was tested against data from 14 meteorological stations around the country, and the correlation coefficient obtained between the fog incidence and the Index was significant beyond the 0.1% level, which 'suggests that the fog index equation produces satisfactory results'.

The Fog Potential Index model represents a first attempt at the prediction of the spatial variation of thick radiation fog. The values of the Index provide coarse estimates for the potential for thick radiation fog at particular locations. The predictions may then be tested against *in situ* observations of visibility. It may well have internal faults which need to be ironed out in the future, but the model cannot be validated or calibrated until data can be obtained from a close network of visibility observation sites, operating over a long period of time in a region of varied topography. If such a network of observation sites could be established, then information on the spatial variability of visibility could be obtained. It may also be improved in the future by conducting visibility traverses (perhaps with a vehicle-mounted mobile transmissometer) along routes where the Fog Potential Index has been evaluated. Airborne thermal mapping (Nixon and Hales 1975; Stove *et al.* 1987) may also assist in the mapping and monitoring of cold-air drainage patterns along particular routes.

6.4 PREVENTIVE AND REMEDIAL ACTION TO DEAL WITH FOG

Fog is a hazard to transport by land, sea and air. Costly adjustments, such as automatic aids for aircraft landing, radar for ships and safety signalling for rail transport, have alleviated the impact of fog in these sectors. Such solutions are, however, largely impractical for road transport. This final section will outline some of the ways in which the hazards of driving in fog may be reduced.

The ideal solution to the fog hazard problem would be to remove or reduce the fog once it has formed; many techniques have been suggested, but all the known techniques are currently too expensive to use on motorways and are not in widespread use. Silverman and Weinstein (1974) provide a through review of fog modification techniques.

The water droplets comprising the fog may be evaporated by heating, for the amount of water vapour which air can hold without condensation occurring and fog forming is dependent upon its temperature. A heating method using oil-burning heaters arranged along the edge of an airfield runway was used during the second World War to disperse fog (known as FIDO or Fog Intensive Dispersal Of), as described by Ogden (1988). Kerosene was burnt in open burners to disperse the fog by heating and evaporation. For this reason the technique is not really applicable to motorways, but this represents the first successful attempt at fog dissipation (and resulted in the successful landing of more than 2000 British aircraft during the war). While such a system is viable over the limited spatial extent of an airfield, the costs of using such a system over stretches of the motorway network would be prohibitive.

An ingenious mechanical method of dispersing fog is the fog broom, first used in Chile to deal with sea fogs which move inland (Bowden 1966). It involves sweeping the fog with a grid of nylon filaments, causing the fog droplets to collect on a light frame attached to the nylon, from where they run down and drip to the ground. The technique has been refined in the United States, where shallow fogs have apparently been successfully cleared from areas of highway in New Jersey employing a rotating grid of nylon filaments.

Most of the techniques employed for fog clearance have involved seeding the fog. If relatively large hygroscopic nuclei, such as sodium chloride (salt) particles are introduced into the atmosphere just prior to fog formation, condensation on the smaller natural nuclei would be inhibited and a smaller number of large droplets would form on condensation and would precipitate out (Highway Research Board

1970). Such seeding has been shown to be effective for some 15 min. Tests in the United Kingdom involving a solution of ammonium nitrate as the seeding agent have proved inconclusive (Stewart 1960). Such chemical methods of fog abatement are numerous, but they tend to be unpopular, partly because their use may have undesirable consequences for the environment.

Perhaps the best possibility of a cheap method of clearing fog is offered by techniques involving the electrical charging of fog droplets positively and negatively in alternate regions of the fog, so that coalescence of the droplets is brought about by electrical attraction and the droplets precipitate out of the fog; laboratory studies on this are currently in progress in the United States. At the moment however there is *no* known means of preventing the occurrence of fog or of dispersing it at an acceptable cost.

A variety of *remedial measures* have been proposed to help drivers to cope with the fog hazard once it has formed. The commonest such measure is the use of overhead lighting in fog-prone localities. In 1986 the United Kingdom had some 2856 km (1775 mi) of motorway, of which only 728 km (455 mi) or 25% were lit (Cairns 1986). The current policy of the Department of Transport is to light all urban motorway sections, all complicated interchanges and areas with high accident rates, but it is under increasing pressure to extend the amount of lighting, despite the cost. The lights most commonly used are sodium lights at a height of 12 m, with a spacing of 35 m. Their function is to illuminate the road and the vehicles using it in all weather conditions, and to show the direction of the road in conditions of fog.

It is doubtful whether such high-level lighting is useful in conditions of fog, however, because of the resultant scattering of light by the fog particles as it travels from the light source to the driver's eyes. This produces a luminous veil over the driver's field of vision, reducing the visual contrast between objects. For an object to be seen under any conditions there must be contrast between it and its surroundings. In conditions of illuminated fog the contrasts of objects are reduced and details disappear; this may prove to be a positive disadvantage to drivers who need to have a sense of depth to their vision in order to perceive motion properly. Under daylight fog conditions, where vehicle outlines generally become visible before rear lights, increasing the level of ambient illumination from such lights has virtually no effect on visibility distances. Under night-time fog conditions, rear lights become visible before vehicle outlines, and while increasing the level of ambient illumination will increase the visibility distances to unlit vehicles and spilt loads, visibility distances to vehicle lights are reduced. Consequently in night-time fogs where vehicle rear lights are the target of interest for the driver, normal 'street lighting' may be regarded as a disadvantage!

It has been shown that the amount of light scattered towards a driver is much less if it travels at right angles to his line of sight, than if he faces the light source (Spencer 1961). A more effective solution to the problem of motorway lighting in fog might therefore be to have continuous *low-level* lighting shining across the road (from the central reservation) at right angles to the driver's line of vision, producing much less scattering towards the driver. Such lighting (as currently used in sections of tunnel) shows the direction the road is taking, thus removing much of the need for the driver in fog to follow the vehicle in front for guidance. Such lighting creates less environmental intrusion away from the motorway, but is more expensive than conventional overhead sodium lighting.

This type of lighting has been tried in a small number of installations in the Netherlands (Moore and Cooper 1972), while a short installation of this type in Pennsylvania has been pronounced a success in fog (Marsh 1957). In Germany a directional type of parapet lighting has been used on a motorway bridge to provide guidance for motorists (OECD 1976). However, the Road Research Programme of the OECD (1971) has concluded about motorway lighting: 'The Group considers it unlikely that artifical lighting can prove a satisfactory solution to the problem of thick fog. It may be desirable therefore to seek a solution based on non-visual means (e.g. electronic aids)'.

The use of more powerful external vehicle lights may be one solution to overcoming the difficulties of seeing and being seen in fog. Increasing the beam intensity of the front lights above 4000 candela (cd) produces little advantage to the driver, although he may see reflecting studs in the road surface better. However, the chances of his vehicle being seen from the front by others are increased. The Road Vehicles (Use of Lights During Daytime) Regulations of 1975 require UK drivers to use headlamps whenever visibility is 'seriously reduced by fog, smoke, heavy rain or spray, dense cloud or any similar condition'. Non-compliance with these conditions carries a substantial fine.

Vehicles are made much more conspicuous in fog at night and by day when they have more powerful lights to the rear. A 1000 cd rear light can be seen from about 60 m compared with the 23 m at which the average rear light is visible in thick fog. The intensity of the rear light should not exceed that of the brake light, so that the latter can be clearly seen if the vehicle's brakes are applied. Lythgoe (1973) has argued that the area of the rear light, as well as its brightness, is important. From October 1979, the Department of Transport has required that all new vehicles are to be fitted with at least one high-intensity rear fog lamp.

Other remedial measures which can be taken once fog or thick fog has developed include the following:

(a) Motorway warning signs, preferably linked to an automatic fog-detecting system. On some motorways a fog-warning system has been integrated into the traffic-control system and matrix signs are used to give information such as FOG, ICE and speed recommendations. Such signs are more effective when accompanied by yellow flashing warning lights. Such systems are in operation in Belgium, West Germany and other parts of western Europe.
(b) Motorway speed restrictions. These are more effective if linked to the word FOG, and serve to alert the driver to the potential hazard ahead. The signs should be switched off as soon as fog improves, otherwise the driver becomes immune to the warning.
(c) Radio broadcasts of the fog hazard on the national motorway network before or at the end of all news summaries and at other fixed times.
(d) Police convoy systems through particularly hazardous stretches of road. A patrol system which goes some way to achieving this has been tried on a heavily used trunk road by the West Yorkshire constabulary (Moore and Cooper 1972).
(e) Internal vehicle warning devices connected to the motorway control system or triggered by sensors beside the motorway to warn of different categories of hazard (fog, accident, roadworks etc.) ahead.

The key to all of these measures is the immediate availability of information on visibility levels from a good automatic fog-detecting system, but it may be some time in the future before this becomes a reality.

The best type of remedial measure is good, careful and responsible driving behaviour. The Fog Code, now officially incorporated into the Highway Code, advises as follows:

(a) Slow down and keep a safe distance to be able to pull up within your range of vision.
(b) Don't hang on to someone else's tail lights; it gives a false sense of security.
(c) Watch your speed; you may be going much faster than you think. Do not speed up to get away from a vehicle that is too close behind you.
(d) Remember if you are driving a heavy vehicle that it may take longer to pull up than the vehicle ahead.
(e) Warning signals are there to help and protect; observe them.

(f) See and be seen; you must use your headlamps or fog lamps and rear lamps.
(g) Check and clean windscreens, lights, reflectors and windows.
(h) If you must drive in fog, allow more time for your journey.

Often this advice is simply not heeded, for 'There may be perfect roads and perfect vehicles, but perfect fools still produce accidents' (West Yorkshire Constabulary 1971).

6.5 CONCLUSION

Fog is that aspect of weather on motorways that drivers fear most. Thick fog tends to reduce traffic volumes and to increase the risk of accidents to those vehicles on the road. There is no known means of preventing the occurrence of fog, or of dispersing it once it has formed at an acceptable cost. Fog is difficult to forecast with accuracy, and the patchy, localized nature of radiation fog makes visibilities particularly difficult to predict. As the hazard cannot be removed, road planners must either attempt to provide better and more effective warning of the hazard through improved visibility monitoring systems, or they must attempt to avoid locating new motorways and roads in locations where the incidence of thick fog is known to be high. The Fog Potential Index perhaps provides a technique whereby such fog black spots may be predicted and identified; it also provides a technique to determine where fog-detecting systems should be located if they are to monitor the worst of the local fog conditions. If drivers observed the recommendations of the Fog Code, the fog hazard would not be eliminated, but its effects in terms of human stress and misery might well be ameliorated.

6.6 REFERENCES

Bowden, D. (1966). A new way of clearing fog. *New Scientist*, **32**, 583–585.
Brown, I. (1970). Motorway crashes in fog – who's to blame? *New Scientist*, **36**, 544–545.
Byers, H.R. (1959). *General Meteorology*. New York: McGraw-Hill.
Cairns, S. (1986). Motorway lighting – Lightec tries to throw some light on our roads. *Surveyor*, **167**, 24.
Caughey, S.J. *et al.* (1978). Acoustic sounding of radiation fog. *Meteorological Magazine*, **101**, 103–113.
Chandler, T. (1965). *The Climate of London*. London: Hutchinson.
Codling, P.J. (1971). Thick fog and its effect on traffic flow and accidents. *Transport & Road Research Laboratory Report 397*. 18pp.

Codling, P.J. (1974). Weather and road accidents. In *Climatic Resources and Economic Activity*, ed. J.A. Taylor. Newton Abbott: David & Charles.

Douglas, H.A. *et al.* (1978). The measurement of fog on motorways. *Meteorological Magazine*, **107**, 242–249.

Findlater, J. (1985). Field investigations of radiation fog formation at outstations. *Meteorological Magazine*, **114**, 187–201.

Highway Research Board (1970). Highway fog. *National Cooperative Highway Research Program No 95*, Washington D.C.

Hogg, W.H. (1965). Climatic factors and choice of site with special reference to horticulture. *Symposia of the Institute of Biology*, 14.

Jack, V.D. (1966). Further work on objective forecasting of visibility. *Meteorological Magazine*, **95**, 114–121.

Jeffery, D.J. & White, M.E. (1981). Fog detection and some effects of fog on motorway traffic. *Traffic Engineering & Control*, **16**, 199–203.

Jiusto, J.E. (1974). Remarks on visibility in fog. *Journal of Applied Meteorology*, **13**, 608–610.

Jiusto, J.E. (1981). Fog structure. In *Clouds – their formation, optical properties and effects* ed. P.V. Hobbs. & A. Deepak. pp. 187–239. New York: Academic Press.

Jiusto, J.E. and Lala, G.G. (1980). Radiation fog formation and dissipation – a case study. *Journal de Recherches Atmospheriques*, **14**, 391–397.

Johnson, H.D. (1973). Motorway accidents in fog and darkness. *Transport & Road Research Laboratory Report 573*. 13pp.

Kocmond, W.C. and Perchonok, K. (1970). Highway fog. *US National Cooperative Highway Research Report 95*.

Lawrence, E.N. (1976). Visibility. In *The Climate of the British Isles*, ed. T.J. Chandler & S. Gregory. pp. 211–223. London: Longman.

Lee, T.F. (1987). Urban clear islands in California Central Valley fog. *Monthly Weather Review*, **115**, 1794–1796.

Lythgoe, J.N. (1973). Countering daylight fog. *Nature*, **243**, 244.

Marsh, C. (1957). Highway visibility in fog. *Illumination Engineering*, **52**, 621–627.

Martin, A. (1974). The influence of a power station on climate – a study of local weather records. *Atmospheric Environment*, **8**, 419–424.

Meetham, A.R. (1964). *Atmospheric Pollution – Its origins and prevention*. Oxford: Pergamon.

Meteorological Office (1969). *Observer's Handbook*. London: HMSO.

Meteorological Office (1985). *The Susceptibility of Fog on the M25 Motorway* Met O 3, Building and Construction Climatology Unit.

Moore, R.L. and Cooper, L. (1972). Fog and road traffic. *Transport & Road Research Laboratory Report 446*. 33pp.

Nixon, P.R. and Hales, T.A. (1975). Observing cold night temperatures of agricultural landscapes with an airplane-mounted radiation thermometer. *Journal of Applied Meteorology*, **14**, 498–505.

Ogden, R.J. (1988). Fog dispersal at airfields. *Weather*, **43**, 20–25 and 34–38.

OECD (1971). *Lighting, Visibility and Accidents*. Paris: OECD. 110pp.

OECD (1976). *Adverse Weather, Reduced Visibility and Road Safety*. Paris: OECD.

OECD (1986). *Road Safety Research 1986*. Paris: OECD. 106pp.
Pedgley, D.E. (1967). Weather in the mountains. *Weather*, **22**, 266–275.
Perry, A.H. (1981). Fog. In *Environmental Hazards in the British Isles*, pp. 79–85. London: George Allen & Unwin.
Pratt, K.A. (1968). The 'Haar' of north-east England. Unpublished B.A. dissertation, Department of Geography, University of Durham.
Prokh, L.Z. (1966). A characterisation of the fogs of the Ukraine. *Trudy UkrNIGMI*, **55**, 43.
Silverman, B.A. and Weinstein, A.I. (1974). Fog. In *Weather and Climate Modification*, ed. W.N. Hess. pp. 355–383. New York: John Wiley.
Spencer, D.E. (1961). Lighting in fog. *Research*, **14**, 55–62.
Stewart, K.H. (1960). Recent work on the artificial dispersal of fog. *Meteorological Magazine*, **89**, 311–319.
Stove, G.C. *et al.* (1987). Airborne thermal mapping for winter highway maintenance using the Barr & Stroud IR18 thermal video frame scanner. *International Journal of Remote Sensing*, **8**, 1077–1078.
Sumner, R. *et al.* (1977). Driving in fog on the M4. *Transport & Road Research Laboratory Supplementary Report 281*. 20pp.
Tanner, J.S. (1952). Effect of weather on traffic flow. *Nature*, **169**, 107–108.
Transport and Road Research Laboratory (1976). Driving in fog on the M4. *TRRL Leaflet LF 632*, 2pp.
Transport and Road Research Laboratory (1987). Fog and road accidents 1985. *TRRL Leaflet 1057*, 2pp.
Unsworth, M.H. *et al.* (1979). The frequency of fog in the Midlands. *Weather*, **34**, 72–77.
Vogel, J.L. and Huff, F.A. (1975). Fog effects resulting from power plant cooling lakes. *Journal of Applied Meteorology*, **14**, 868–872.
West Yorkshire Constabulary (1971). *Accident prevention during conditions of fog on the A1 in West Yorkshire*. West Yorkshire Constabulary internal report.
Wheeler, D. (1986). A study of sea fret – 27 April 1986. *Journal of Meteorology*, **11**, 311–317.
White, M.E. and Jeffery, D.J. (1980). Some aspects of motorway traffic behaviour in fog. *Transport & Road Research Laboratory Report 958*, 12pp.
Willett, H.C. (1928). Fog and haze. *Monthly Weather Review*, **56**, 435.

Chapter Seven

Blowing dust and highways: the case of Arizona, USA

A.I. Brazel

Blowing sand and dust produce hazardous driving conditions along major thoroughfares in portions of the Southwest of the USA. Several fatal accidents have resulted on local, state, and interstate highways in the region. This chapter details some of the problems of blowing dust on Arizona's highway network. Arizona dust studies started at an applied level (Marcus 1976) and, subsequently, fundamental research on dust emission and factors of dust generation took place (Brazel and Nickling 1987).

This chapter is written as a retrospective analysis of the highway dust problem, since fundamental information is now available on dust generation and climate relationships from recent field data and analyses of the meteorological record. Sections 7.1 and 7.2 discuss meteorology and climatology, and factors of dust emission and dust generation, followed by a brief presentation (sections 7.3 and 7.4) of the transportation dust hazard and dust warning system developed on the Arizona highway network. Future dust problems on the highways are mentioned in section 7.5. The future scenarios are important in the light of recent severe dust accidents on the highways and climatologists' recent analysis of CO_2 scenarios and drought potentials. Two decades ago the state of the art in scenario work would not allow for such future glimpses of the climate system, and, in this case, possible future dust problems.

The Arizona Department of Transportation (ADOT) in recent years had thought that the dust problem had gone away. As a result, it ended dust-warning signs along a historically dust-prone stretch of the interstate highway system in central Arizona, and cut back programmes of education designed to provide driver awareness of the dust storms that frequent the state in the summer months. However,

Figure 7.1 Haboob-like dust wall caused by large down draught in a thunderstorm passing through Phoenix, Arizona.

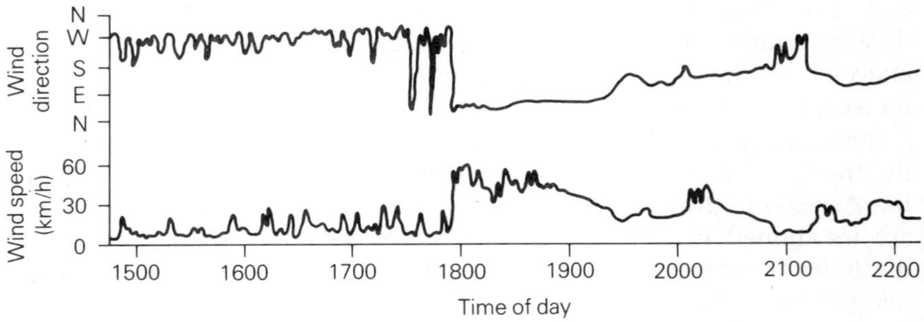

Figure 7.2 Sample wind trace during a severe blowing-dust event similar to dust wall passage shown in Figure 7.1.

the period 27–29 August 1988 was a grim reminder that the problem still exists. During this time two major fatal pile-ups of vehicles occurred on the interstate system between Phoenix and Tucson, Arizona. As will be mentioned later, long-term changes of the dust source areas (changes which are subtle and may go unnoticed) provide a major key to understanding the blowing dust hazard from year to year in central Arizona.

7.1 METEOROLOGY AND CLIMATOLOGY OF THE SOUTH-WEST AND ARIZONA

Arizona and the south-western United States experience dust storms all year round. The most severe and intense storms are associated with high winds generated by thunderstorms and strong convective activity in summer. This activity is part of the 'Arizona Monsoon' season, normally lasting from July until September (Carleton 1987). Figure 7.1 illustrates the classic 'haboob'-like dust wall emanating from a thunderstorm cell moving across the Phoenix metropolitan area. A typical wind trace is shown in Figure 7.2. The passage of the dust wall, microburst and/or gust front accelerates wind to speeds far exceeding the thresholds necessary to entrain dust from many of the various land types in the southern Arizona region. In the example shown, at 1800 hours Arizona time, winds increased from 12 km/h (7.4 mile/h or 3.3 m/s) to 60 km/h (37.2 mile/h or 16.7 m/s) in a matter of minutes. Light westerly winds changed to strong wind from an easterly direction: the most common direction associated with large-scale convective meso-scale system movement in southern Arizona during summer. These rapid changes are due to local-scale convective activity, usually occur in late afternoon/early evening, and are difficult to predict. They promote virtually blinding walls of dust across major highways of southern Arizona and represent an extremely dangerous hazard.

Climatological and meteorological analyses of the dust problem have been completed by a number of researchers over the last decade (Pewe 1981; Goudie 1983; Nickling and Gillies 1986). Figures 7.3 and 7.4 and Tables 7.1 and 7.2 present general findings of a dust-frequency analysis for the South-West and particularly for Arizona (after Changery 1983; Brazel and Nickling 1986 and 1987). There are few sites for which historical dust frequencies can be determined. For the Mojave and Sonoran Desert regions, there are only sixteen sites with sufficient data to develop a dust climatology (Figure 7.3). A seventeenth site, Tonopah, Nevada, is included for comparative purposes with Great Basin dust-event statistics.

In the meteorological records, a dust event is recorded when the prevailing visibility around the weather station drops below 11.6 km (7 miles). A weather observer indicates the beginning of this obstruction to vision (when visibility drops below 11.6 km) and the ending time,

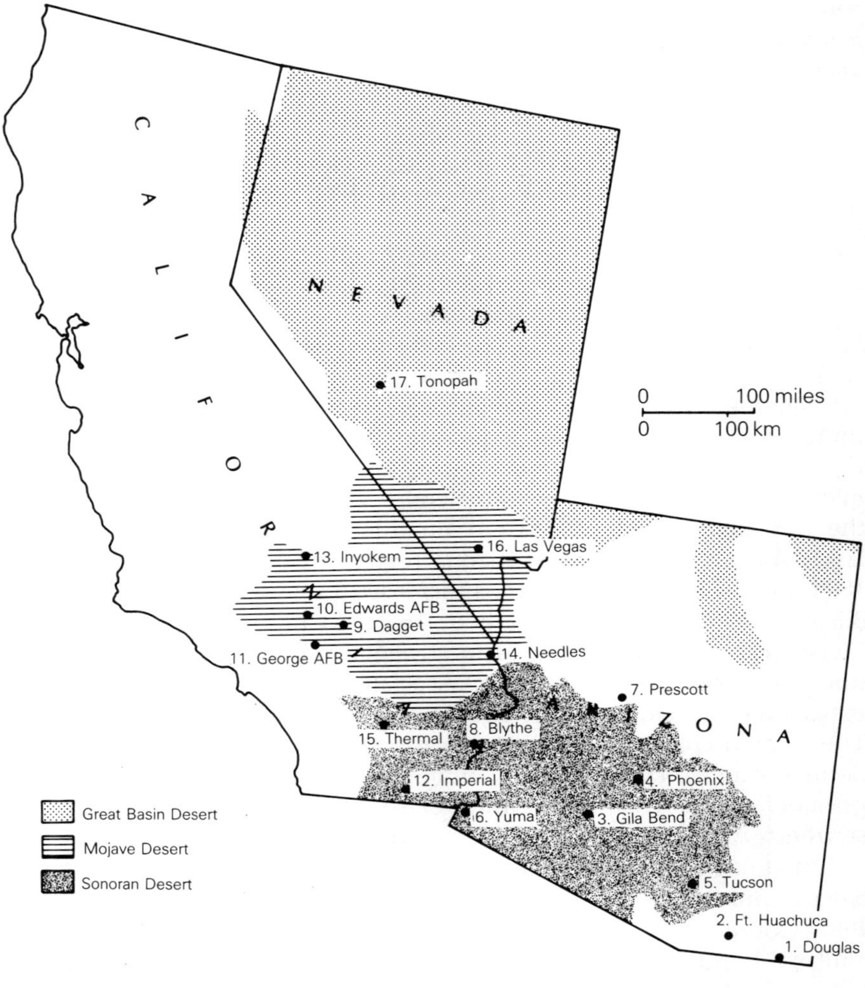

Figure 7.3 Weather stations which record visibility reductions due to blowing dust and sand in the south-west USA.

when visibility exceeds the 11.6 km value. The observer also logs the lowest visibility during the event. More severe dust events, labelled dust storms, occur when the visibility drops below 1 km, the international definition of a dust storm (Goudie 1978). Further remarks are usually entered on standard weather-observation sheets, such as the quadrant of the storm, wind speeds over the duration of the event, cloud conditions, temperatures, and dew points. These data are helpful in resolving where storms come from and the severity of associated atmospheric synoptic conditions which promote blowing dust. In retrospect, analyses of dust sources are possible to some degree with the meteorological data records from major weather stations of a region. The secondary and tertiary stations of weather networks do not usually record dust events, and thus historical information is limited in resolving the location and severity of blowing dust.

Blowing dust can occur in the desert with visibilities generally exceeding 11.6 km. These events are not normally logged in weather records. Thus, dust-frequency data are limited to the more severe visibility reductions. However, these data are appropriate for analysis of the highway dust problem, because the mild events do not usually restrict drivers' visibility nor cause major accidents on the road network of Arizona. For other problems, such as air quality, human respiratory problems, climatic change and desertification, a spectrum of all dust conditions would generally be required.

Tables 7.1 and 7.2 list general climatic conditions of the seventeen sites shown in Figure 7.1. Elevations range from −20 to 1679 m, precipitation ranges from 64 to 479 mm; daytime maximum temperatures range from 18.9 °C to 31.6 °C; and pan evaporation ranges from 940 to 1778 mm. The lower range of annual precipitation (64 mm) fits the category suggested by Goudie (1978) below which few dust storms would occur. Goudie's reasoning is that a synoptic event (perhaps with some subsequent rainfall) is necessary in the first place to cause enough wind for dust entrainment. For the region of the South-West, the upper value exceeding 400 mm is a limit of no recorded dust events in a year. Notice in Table 7.2, for example, that Prescott, Arizona records very few dust events per year. In addition to the higher precipitation, ground vegetative cover restricts wind erosion (in this case a Ponderosa Pine forest cover).

Table 7.1 also includes the probability of thunderstorms, in percent, associated with dust-event occurrences. For sites in the Mojave and Great Basin Deserts, dust events are associated with strong topographic channelling of winds during winter storms with associated steep pressure gradients. In Arizona, there is a good probability that thunderstorm activity is coincident with blowing dust events. The spatial differences across the seventeen-site network and the seasonal timing of the events result from a combination of initial weather disturbance influences and physical characteristics of local surfaces around each site. Blowing dust incidences across the seventeen-site network range from 0.5 to 35.6 events per year (<11.6 km visibility) and 0.1 to 5.1 events per year (<1 km visibility). These values are considerably

Table 7.1 *South-western United States climate summary for the Sonoran–Mojave desert region*

Station	ELEV. (m)	Precipitation† (mm)			Temp. (°C)		EVAP (mm)	C (%)	T. storm (%) prob‡
		S	W	A	Max	Min			
(1) Douglas	1249	204	105	309	25.4	7.4	1143	48	25
(2) Ft. Huachuca	1422	222	101	323	23.9	9.7	1016	18	13
(3) Gila Bend	224	56	92	148	31.6	13.4	1143	283	35
(4) Phoenix	340	66	115	181	29.5	14.1	1143	88	34
(5) Tucson	788	153	125	278	27.6	12.3	1067	143	41
(6) Yuma	63	21	46	67	31.0	15.4	1397	485	7
(7) Prescott	1679	219	260	479	20.4	2.9	940	10	27
(8) Blythe	120	31	55	86	30.6	15.3	1778	502	12
(9) Daggett	584	30	67	97	24.0	9.3	1448	975	3
(10) Edwards AFB	702	5	135	140	24.4	7.8	1397	172	2
(11) George AFB	879	0	129	129	23.9	8.9	1524	225	4
(12) Imperial	−20	18	46	64	30.9	14.7	1270	445	3
(13) Inyokern	699	10	93	103	26.1	10.0	1270	312	1
(14) Needles	278	39	73	112	29.7	15.4	1651	135	18
(15) Thermal	−34	18	54	72	31.1	13.7	1397	483	2
(16) Las Vegas	659	38	68	106	26.2	11.3	1346	573	10
(17) Tonopah	1654	50	74	124	18.9	2.2	1016	140	4

† Data from U.S. Department of Commerce, U.S. Air Force Weather Service and Brazel and Hsu (1981). S = summer (June–Sept); W = winter (October–May); A = Annual.
‡ Thunderstorm probabilites from Changery (1983).

less than the annual frequencies in the Sudan, many areas of the Middle East (e.g. Kuwait, Saudi Arabia), southern USSR, Afghanistan, northern China, portions of the Sahara desert, and the Southern Great Plains of the USA (Goudie 1983).

Reasons for lower values relate to substantial vegetation cover and/or fewer storms in the south-western USA, depending on which site in the network is analyzed. The ratio in the third column of Table 7.2 consists of the number of severe events (visibility below 1 km) to all events which caused visibility to register below 11.6 km. This ratio is not constant across the network, but varies from 0.06 to 0.27. Thus, in the case of Ft Huachuca, Las Vegas, Edwards AFB, Blythe, George AFB, and Inyokern there are many severe events in relation to the total number of events experienced at these sites. In comparison, Douglas, Phoenix, Tucson, Prescott, Needles, and Tonopah register many more

Table 7.2 *South-western United States dust event summary*[†]

Station	*Frequency per year*		
	11.3 km	*1 km*	*Ratio (1 km/11.3 km)*
(1) Douglas	1.8	0.1	0.06
(2) Ft. Huachuca	0.9	0.3	0.33
(3) Gila Bend	10.8	2.1	0.19
(4) Phoenix	18.1	1.6	0.08
(5) Tucson	3.8	0.3	0.08
(6) Yuma	23.9	4.7	0.19
(7) Prescott	0.5	0.03	0.06
(8) Blythe	12.7	3.0	0.23
(9) Daggett	9.1	1.2	0.13
(10) Edwards AFB	20.3	5.0	0.25
(11) George AFB	13.0	3.0	0.23
(12) Imperial	22.8	2.5	0.11
(13) Inyokern	8.0	1.8	0.22
(14) Needles	2.8	0.2	0.07
(15) Thermal	35.6	5.1	0.14
(16) Las Vegas	13.2	3.6	0.27
(17) Tonopah	2.0	0.2	0.10
Mean	12.0	2.1	0.16

[†] Data from Changery (1983)

mild dust events and very few severe ones. It is difficult to resolve why these differences in the severity ratio exist across the region. However, the reasons most probably relate to differences in ground cover as well as the severity of synoptic events. Research on this question will greatly assist in interpretations of dust hazards on highways, since ground cover can potentially be manipulated; the severity of the weather is beyond our control at present.

Another important consideration of regional dust climatology is the timing and duration of blowing-dust events. Figure 7.4 illustrates the percentage of blowing-dust events per month at three sites representing a transect from the winter precipitation regime of Southern California to the summer monsoon precipitation zone of Southern Arizona. Most dust events occur during spring in the west. They take on a bimodal distrubution in the transition area (spring and summer), and there is a summer concentration of dust events in the eastern, Arizona monsoon location. All three areas experience spring peaks in

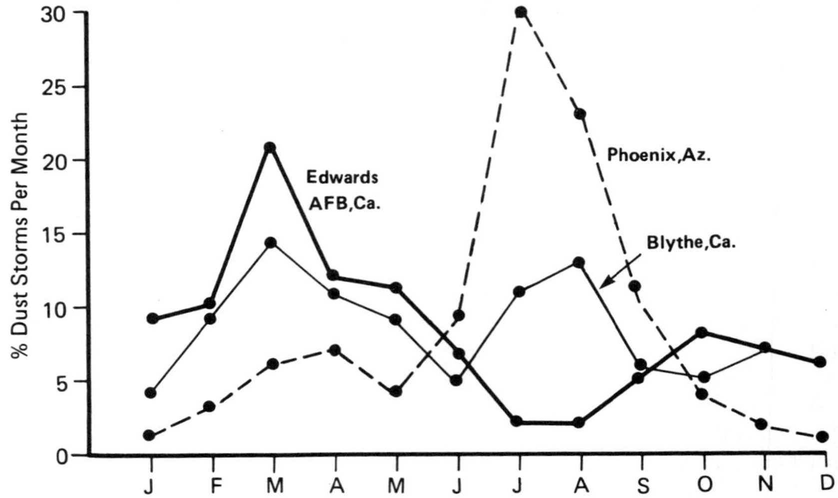

Figure 7.4 Percentage blowing dust by month (visibility conditions less than 11.3 km) in a transect across the south-western USA.

dust events, related to expansion of the circumpolar vortex in late winter in the northern hemisphere and subsequent frontal activity across the American South-West (Orgill and Sehmel 1976). The summer peaks of the central and eastern sites are related to south-western monsoon mesoscale convective system activity. For three Arizona sites (Table 7.3), there are subtle differences in the percentage of dust events per month and the total time (duration) of blowing dust each month. These differences are a function of site attributes as well as general synoptic influences of spring and summer. Tucson has very few dust events compared to Phoenix, yet Tucson is closer to major source regions of mesoscale system development in summer (Hales 1974). There is a considerable dust-event frequency gradient between Tucson and Phoenix (a most important highway region of blowing dust).

Dust events can be traced in a general way to specific synoptic types. Brazel and Nickling (1986) analysed four synoptic situations which are associated with major blowing-dust events at various sites in Arizona. Table 7.4 illustrates the relationship of blowing-dust incidences (and their characteristics) with synoptic weather types. The same three southern Arizona sites (Yuma, 64 m; Phoenix, 339 m; and Tucson, 780 m) are analysed in addition to a site not in the Sonoran desert region of Figure 7.3, but in the higher-elevation, semi-arid, scrub-to-

Table 7.3 *Seasonal dust, percentage per month and percentage time observed for Yuma, Phoenix, and Tucson, Arizona, (1965–1980)*

Month	*Yuma*		*Phoenix*		*Tucson*	
	% per month	*% time blowing dust*	*% per month*	*% time blowing dust*	*% per month*	*% time blowing dust*
J	3	.12	0	.00	0	.00
F	3	.11	1	.02	0	.00
M	6	.18	1	.01	0	.00
A	17	.28	3	.13	0	.00
M	6	.17	6	.03	9	.01
J	3	.03	8	.05	37	.01
J	18	.08	28	.15	27	.02
A	27	.17	35	.20	27	.01
S	12	.21	15	.15	0	.00
O	0	.00	2	.00	0	.00
N	2	.03	1	.01	0	.00
D	3	.10	0	.00	0	.00

* Relative to less than 1.6 km visibility obstruction

grass desert of the Colorado Plateau. This location is Winslow, Arizona (1490 m).

The four general synoptic types associated with blowing dust at these sites are: (a) frontal zones of a mid-latitude cyclone; (b) thunderstorm, convective activity; (c) tropical disturbances; and (d) upper-level lows and cut-off lows. The frontal type is subdivided into two subtypes; *pre-frontal dust* and *post-frontal dust*. The pre-frontal type, Type 1A, occurs when most of the dust blows across the surface as a cold front passes by the location. The post-frontal type, Type 1B, occurs when dust is stirred up around a location after the surface front has passed by a site. The post-frontal type occurs when high pressure builds in rapidly behind the cold front and high winds are sustained for a lengthy period of time. This latter type is quite frequent in the Great Plains region of the USA (Gillette, in Pewe 1981). Next to thunderstorm influences, this type is most frequent in Southern Arizona and on the Colorado Plateau.

The *thunderstorm/convective* type, Type 2, can be highly localized in origin and occurs in summer. At present there is no classification of the wide range of spatial and temporal conditions associated with this type for the Arizona region. Much research needs to be developed on

Table 7.4 *Synoptic and event characteristics of blowing dust for Yuma, Phoenix, Tucson, and Winslow, Arizona (1965–1980) (after Brazel and Nickling 1986)*

Type	*Yuma*					*Phoenix*					*Tucson*					*Winslow*				
	n	*Time (h)*	*Direction (degrees)*	*ΔTime (min)*	*Duration (min)*	*n*	*Time (h)*	*Direction (degrees)*	*ΔTime (min)*	*Duration (min)*	*n*	*Time (h)*	*Direction (degrees)*	*ΔTime (min)*	*Duration (min)*	*n*	*Time (h)*	*Direction (degrees)*	*ΔTime (min)*	*Duration (min)*
Intense dust storms																				
1A	9	1325*	250	140	208	2	1389	260	42	78	0	–	–	–	–	1	0858	190	28	140
		(410)†	(142)	(117)	(172)															
1B	13	1622	291	299	428	5	1534	204	132	229	0	–	–	–	–	10	1159	223	124	240
		(378)	(20)	(190)	(254)		(342)	(91)	(102)	(220)										
2	25	1439	140	9	75	57	1834	129	9	54	7	1813	106	4	17	6	1616	165	4	31
		(748)	(87)	(14)	(63)		(431)	(79)	(19)	(37)		(144)	(46)	(4)	(17)		(169)	(99)	(5)	(27)
3	2	1439	110	16	258	2	1847	150	26	75	2	1715	110	14	13	0	–	–	–	–
4	2	1459	240	125	218	5	1708	238	23	32	2	1530	110	9	12	0	–	–	–	–
							(136)	(103)	(9)	(14)										
	51					71					11					17				
Moderate to weak dust storms																				
1A	30	1427	283	–	138	1	1646	340	–	54	0	–	–	–	–	2	1315	160	–	373
		(512)	(110)		(102)															
1B	51	1228	283	–	248	9	1507	260	–	135	4	1118	233	–	261	11	1138	236	–	239
		(675)	(50)	–	(197)		(672)	(35)		(82)							(495)	(49)		(191)
2	19	1122	167	–	106	28	1637	167	–	57	0	–	–	–	–	0	–	–	–	–
		(927)	(70)		(70)		(703)	(87)		(72)										
3	1	2140	260	–	95	0	–	–	–	–	0	–	–	–	–	0	–	–	–	–
4	12	1619	270	–	214	3	1631	227	–	103	0	–	–	–	–	0	–	–	–	–
		(782)	(56)		(84)															–
	113					41					4					13				

*Mean
† Standard deviation where appropriate

this topic and is currently underway among Arizona universities, the National Weather Service, and the National Severe Storms Laboratory of NOAA. Several previous individual thunderstorm and blowing-dust event descriptions point to the great complexity associated with this type (Idso *et al.* 1972; Ingram 1972; Hales 1975, 1977; and Brazel and Hsu 1981).

The *Tropical disturbance* type, Type 3, is associated with a tropical storm and unstable, moist air from decaying stages of the storm (Fors 1977). This type usually occurs in late summer and fall, and affects southern Arizona. However, these events are relatively rare.

The *upper-level low*, Type 4, is a high-amplitude low-pressure trough which often develops a cut-off circulation over the South-West with much instability (Douglas 1974). It may reside over the region for several days and promote strong surface winds, usually occurring in spring or fall months.

Table 7.4 details the frequency of these events and their association with blowing-dust conditions. In Table 7.4 frequencies over the period 1965–1980 are listed in two categories; intense dust storms (in this case arbitrarily assigned a visibility less than 1.6 km), and moderate-to-weak dust storms (visibility between 1.6 and 11.3 km). In addition, the following are presented: (a) the time of day when the dust event started (time); (b) the average 10 m wind direction during the dust storm (direction); (c) the storm duration or length of time below 11.3 km visibility (duration); and (d) the time for visibility to decrease from 11.3 to 1.6 km during the intense events (delta time). These parameters are coded by synoptic type.

For Type 2, the thunderstorm type, storm directions are typically from the South-East across southern Arizona and occur during late afternoon and early evening. Frontal and upper-level trough-induced dust is predominantly from northerly and westerly directions. Wind speeds associated with dust events are quite variable and in Arizona range from 4 m/s to over 22 m/s (see Figure 7.5). For comparative purposes, several north-west African sites are averaged and wind-speed data are also shown in Figure 7.5 (after Helgren and Prospero 1987). In Arizona, higher wind speeds are required for dust generation than in the more arid areas such as the north-west African sites.

7.2 DUST-GENERATION FACTORS

In addition to atmospheric influences, other factors affect the generation of blowing dust within a region. These include: soil type, vegetative cover, frequency and magnitude of surface disturbances,

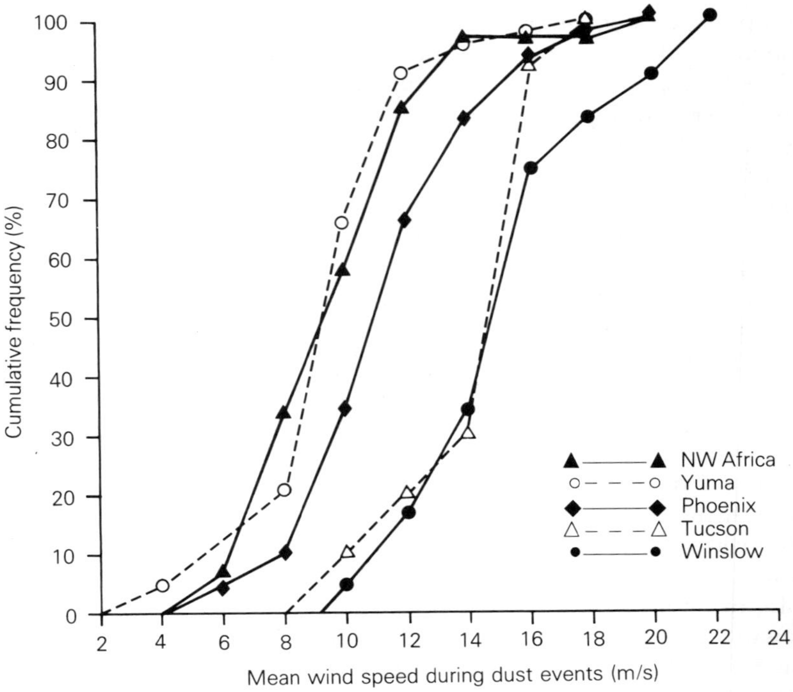

Figure 7.5 Mean wind speed (m/s) cumulative frequency distribution for dust events in NW Africa and Arizona.

real-time and antecedent soil moisture and precipitation, and the human use of land areas suspected to be dust sources. Dust sources are not necessarily widespread in nature, but may consist of small point sources (e.g. a cleared field or an alluvial surface) and line sources (e.g. dirt roadways and river channels) which in the aggregate provide considerable dust for entrainment by major weather systems (Nickling and Gillies 1986; Nickling and Gillies, in press). Figures 7.6(a) and (b) show the generalized soil types and vegetation types for Arizona.

Arizona is not entirely an arid state. The lower elevations of the north-eastern part of the state, on the Colorado Plateau, are arid. Soil conditions are classified as desert types with shrubs and some grasses on the surface. The south-western third of the state is the most arid zone of the region. Sparse creosote bush dots the light-coloured desert soils of the south-western region. Exceptions are deeper soils found

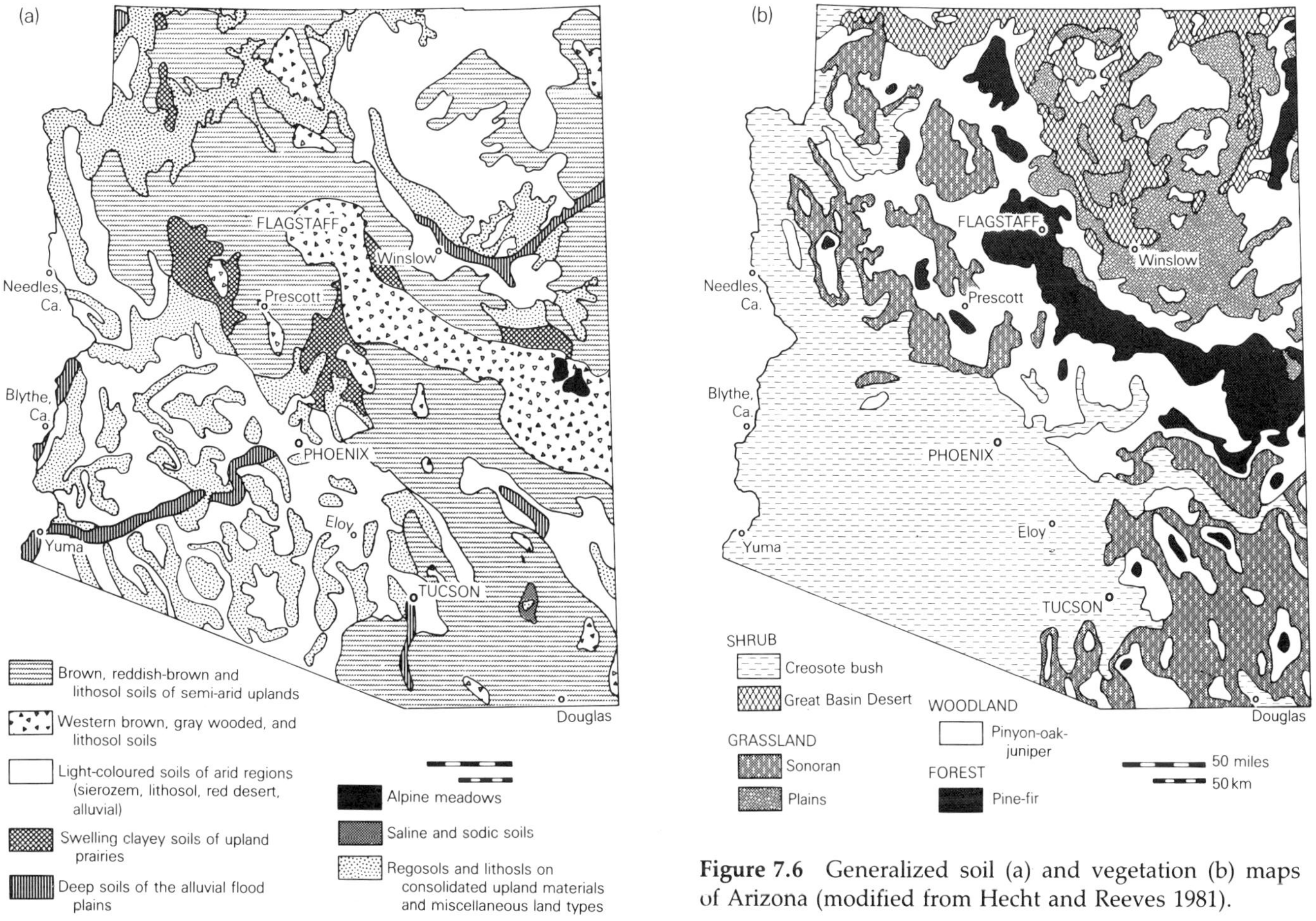

Figure 7.6 Generalized soil (a) and vegetation (b) maps of Arizona (modified from Hecht and Reeves 1981).

in alluvial flood plains (e.g. near Tucson and between Yuma and Phoenix). A major portion of the central and south-eastern part of Arizona is covered by lush grasslands and/or forest cover.

Thus, the upland environment represented by Winslow and the Tucson, Phoenix, Yuma triangle is the most susceptible to blowing dust. A large fraction (about 40%) of surface sediment material in these areas is finer than 0.074 mm diameter (silt) on desert surfaces. The fraction is even larger on alluvial and retired farmland surfaces (Marcus 1976).

Human use of the land can significantly alter the potential for blowing dust in an area, in addition to the natural factors that control dust generation (e.g. Wilshire *et al.* in Pewe 1981). Figures 7.7(a), (b), and (c) illustrate cropland acreages in 1974, remote subdivision development areas in 1974, and ground water draw-down (between 1940 and 1972). Along the Yuma, Phoenix, and Tucson triangle, significant agricultural cropland has been developed in lowland riverine environments. Through time, ground-water pumping for agriculture has drastically dropped the water table, so that in many areas – particularly between Phoenix and Tucson – much of the land has been abandoned from agricultural use and left in speculation for land development of subdivisions as well as for other uses. This land type has caused local aggravation of the blowing-dust problem near highways.

Natural factors of local dust generation include the threshold wind speed required for entrainment, antecedent moisture, vegetation invasion, and surface-crusting processes of the soil. Table 7.5 lists wind-threshold values for various types of surface in the South-West (after Clements *et al.* 1963 and Nickling and Gillies 1986). A large range of 5.1 m/s to over 16.0 m/s occurs. Mine tailings, abandoned land, and disturbed desert surfaces (e.g. by cattle grazing and off-road vehicle use) have low wind thresholds and are more susceptible to generation of blowing dust. Natural undisturbed desert surfaces have high wind thresholds and do not supply much dust to the atmosphere except during infrequent, severe atmospheric events. One major reason is that surface 'crusting' is typical of these surfaces, unless they are disturbed (see Figure 7.8). Crusting is widespread over soil surfaces of the desert and has been analysed in detail by Gillette *et al.* (1982). A subtle relationship has been noted between moisture and the survival of this crusting process. Too much intense rain of short duration with subsequent drying may actually assist in breaking up the surface crust, exposing loose soil from below to further wind effects. Light rains over a long period may provide ripe conditions for crust formation and maintenance. However, in areas where there is enough winter or summer rainfall, perennial vegetation cover invades the desert surface

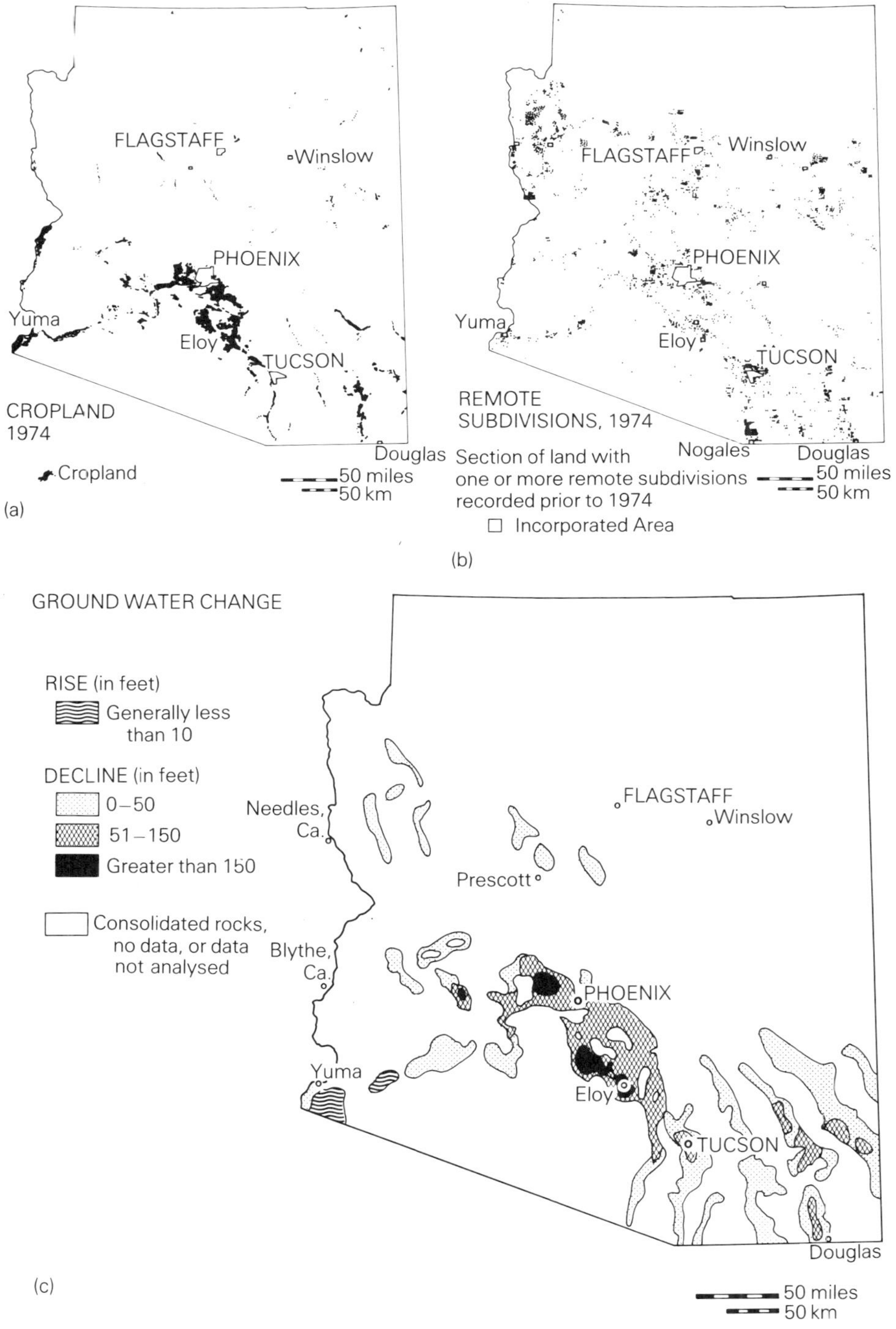

Figure 7.7 (a) cropland, 1974, (b) remote subdivisions, 1974, and (c) ground water change (1940–1972) in Arizona (modified after Hecht and Reeves 1981). (1 ft = 0.305 m)

Table 7.5 *Wind threshold values for type surfaces in the South West (after Clements* et al. *1963 and Nickling and Gillies 1986)*

Surface type	*Threshold speed (m/s)*
Mine tailings	5.1
River channel	6.7
Abandoned land	7.8
Desert pavement, partly formed	8.0
Disturbed desert	8.1
Alluvial fan, loose	9.0
Dry wash	10.0
Desert flat, partly vegetated	11.0
Scrub desert	11.3
Playa (dry lake), undisturbed	15.0
Agriculture	15.6
Alluvial fan, crusted	16.0
Desert pavement, mature	>16.0

Figure 7.8 Typical armoured desert crust on a barren, abandoned agricultural field in central Arizona near Eloy, Arizona (ice axe for scale).

and further affects crusting in addition to raising the wind threshold for dust generation (e.g. Lougeay *et al.* 1987).

This latter factor appears to be very important in the Arizona case. during years of high winter rainfall in southern Arizona, the subsequent spring and summer frequencies of blowing dust are lower. Field investigations of many of the surface types in the desert (particularly abandoned fields) during dry winters and wet winters has revealed over 60% difference in secondary vegetation succession and in perennial vegetative grass cover on the surface (Karpiscak 1980). This cover typically grows up from February through to April and may survive into the monsoon rainy period of summer, even if only in stubble form on the surface. Karpiscak (1980) studied in detail the invasion by vegetation (grasses, weeds and bushes) of abandoned agricultural fields and natural desert surfaces in the central Arizona region near Casa Grande and Eloy: the critical dust-generation zone where many traffic accidents have occurred. He identified over 160 plant types that have colonized surfaces of the region after these surfaces have been cleared. Several exotic and native grasses and weeds grow in response to critical timing of moisture and temperature in the winter period. Their distribution can be widespread on desert surfaces. In Figures 9(a) and (b), note the contrast in vegetation coverage. The barren surface photograph was taken during a year of little winter rainfall. The dark, almost continuous mat of vegetation shown in Figure 9(b) developed during a wet winter in central Arizona. Lougeay *et al.* (1987) have identified the spatial extent of this vegetation difference for two contrastingly wet and dry winters.

The distribution of vegetation on fields and desert surfaces is a complex process depending on seed source, moisture, grazing activity, other land disturbances such as off-road vehicle effects, and other recreational use of idle lands in the area (Karpiscak 1980). Plants that create widespread coverage and respond to timing of winter moisture are numerous. Important plants are *Schismus arabicus, Erodium cicutarium*, and *Plantiqo insularis* (Karpiscak 1980).

Table 7.6 shows correlations of a winter soil-moisture indicator with annual dust frequencies for the ten selected stations shown in Figure 7.3. This indicator is the *Palmer Drought Severity Index*, which has extensive use in climatology and soil moisture analysis (Palmer 1965; Karl and Knight 1985). These correlations, which are significant for Yuma, Blythe, Las Vegas, and Phoenix (but not for the southern California sites), show indirectly the effect of the vegetative control on subsequent spring and summer dust generation (Brazel and Nickling 1987). The relationship is most important to note as we review the retrospective analysis of the highway dust problem later on.

(a)

(b)

Figure 7.9 (a) Vegetation invasion on barren desert terrain during a wet winter, and (b) barren surfaces devoid of vegetation during a winter drought year. These sites are typical of widespread conditions in central Arizona between Phoenix and Tucson, particularly on abandoned agricultural surfaces.

Table 7.6 *Correlation of PDSI (November–April) with annual dust frequencies (January–December) for ten selected stations in Figure 7.1. The top four places, all in the central and eastern portions of the region, attain significant* r^2 *values at the 95% confidence level. All others show no statistical significance at the 95% confidence level*

Site	r^2 *values*
Yuma	0.38
Phoenix	0.36
Blythe	0.26
Las Vegas	0.21
Inyokern	0.14
Edwards	0.08
Daggett	0.08
Imperial	0.06
George	0.04
Thermal	0.00

7.3 THE HIGHWAY HAZARD

Major highway systems in Arizona are relatively new in comparison with the rest of the country. Most have been constructed only within the last three decades. Interstates 8, 10, 17, and 40 are major transportation routes through dry portions of the state (see Figure 7.10(a)). The heaviest traffic flow year-round is between the two largest cities in the state: Phoenix (ca. 1.8 million in the metropolitan area) and Tucson (ca. 0.3 million). The connecting I-10 and portions of I-8 have witnessed more dust-related traffic accidents, by far, than any other location in the state (see Figure 7.10(b)). This map shows the years 1969–1970, just after the highway was put in between the two cities. A major cluster of dust-related accidents occurred along this stretch of the transportation system. Subsequent yearly dust-accident geographic patterns almost always resemble that shown in Figure 7.10(b). Figure 7.11 shows the percentage frequency of occurrence by hour and by month of dust-related traffic accidents along various milepost distances of I-10 between Phoenix and Tucson. Also shown is the frequency of thunderstorms per hour and per month from a Tucson weather site: Davis Monthan Air Force Base. The correspondence between dust

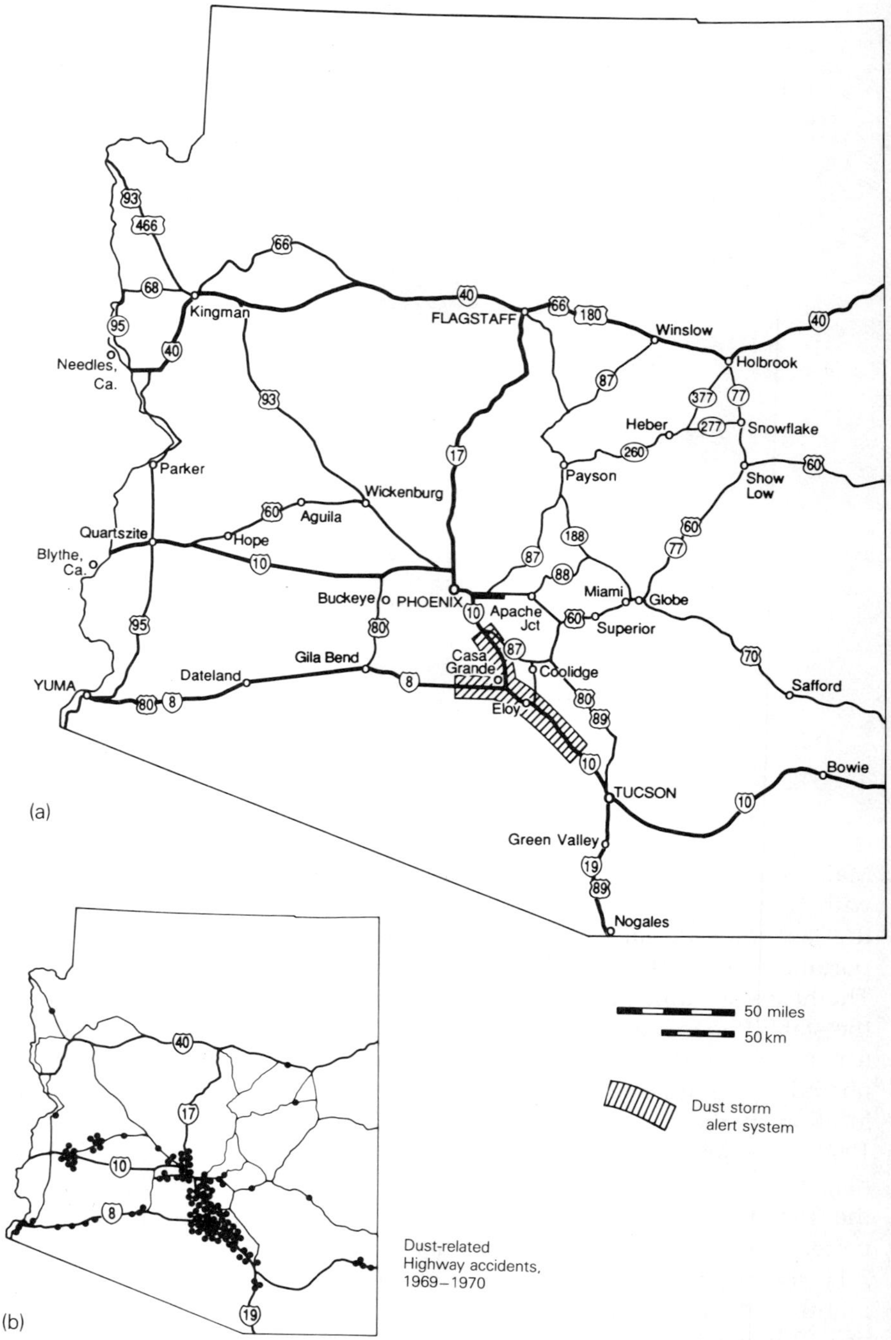

Figure 7.10 (a) Major highways in Arizona showing the dust-storm alert system location, and (b) 1969–1970 dust-related highway accident locations. Each dot is one accident event, which could consist of more than one vehicle (after ADOT 1976).

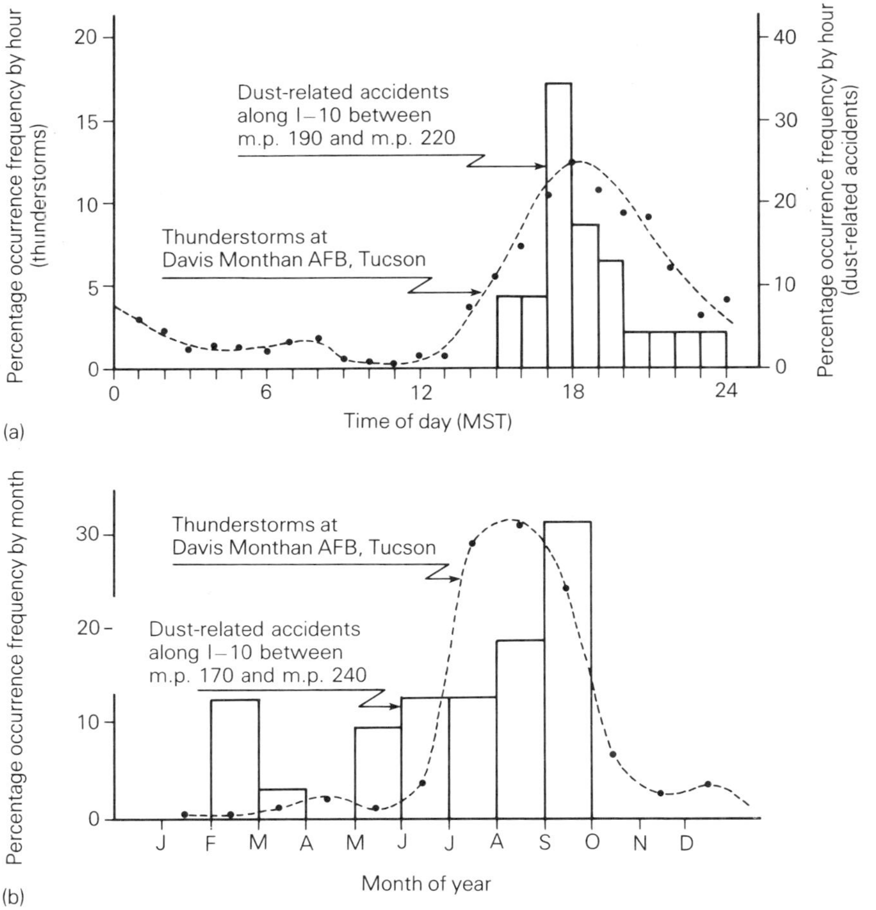

Figure 7.11 (a) Diurnal occurrences of thunderstorms and dust accidents along a stretch of I-10 in central Arizona; (b) monthly frequencies of thunderstorms and dust accidents along I-10 in central Arizona (after Brazel and Hsu, in Pewe 1981).

accidents and the thunderstorm frequency is striking and relates to the discussion above on synoptic influences on dust-storm generation.

The photograph in Figure 7.12 is a stark reminder that accidents do not usually involve just one automobile, but typically many vehicles in a single or multiple chain-reaction accident. Blinding dust walls blot out visibility in a leading vehicle, which either stops suddenly or runs off the road with other vehicles closely following suit. Often a semi-

Figure 7.12 Major dust accident in September 1975 along I-10 involving several trucks and automobiles (after Pewe 1981).

Figure 7.13 Radio-controlled roadside dust-warning sign maintained by Arizona Department of Transportation. Note radio antennae in upper right (after ADOT 1976).

trailer truck with better visibility conditions, the driver being well above the immediate highway surface, triggers accidents by maintaining faster speeds through the dust wall. The photograph shows several trucks and many cars involved in one multi-vehicle accident event.

The dust traffic accidents that followed immediately after interstate highway construction caused a study of this hazard by the Arizona Department of Transportation (Burritt and Hyers, in Pewe 1981). The target area for study and warning systems is shown as a Y-shaped cross-hatched area in Figure 7.10(a). A dust warning system was initiated prior to 1972, when radio-controlled warning signs were positioned on I-8 and I-10, 40 signs being placed along a 31-mile (50 km) stretch near Eloy and Casa Grande, Arizona (see photograph in Figure 7.13). These signs feature flashing lights to draw attention to different warning messages, depending on the anticipated severity of the storm. The weather service transmits forecast severity to Department of Safety personnel in the field near the highway. The system is on longer operational. ADOT determined that it did not reduce accidents, because during storms dust prevented motorists from seeing the warning signs when in the middle of the storm. The system cost $800 000 and yearly maintenance costs exceeded $20 000. In addition to the ADOT in-house studies which showed signs were ineffective, state budgetary problems precluded continuance of the system.

Although ADOT officials claim that when motorists are in the dust wall, they can not see the signs, and hence they are useless, it seems logical that the warnings should have been conveyed long before this point. If dust were reported and some signs were outside the blinding dust locale, they still should be effective as a warning device. Several statistical studies indicate no effect of signs on dust-accident rates in central Arizona. The retrospective analysis section below suggests that climate change significantly affects the accident rate in addition to warning-system development.

7.4 RETROSPECTIVE ANALYSIS OF DUST STORMS, ACCIDENTS AND CLIMATE

The first evaluation of the effectiveness of the dust warning system (DWS) was conducted in 1975. The system was designed so that each of the 40 signs on I-10 and I-8 between Phoenix and Tucson was controlled individually. The system was modified in 1976. ADOT field personnel obtained two sets of data: (a) recordings of the mode displayed on the warning signs, estimations of driver speeds, and whether the displayed sign mode was appropriate to local conditions; and (b) recording of numbers of vehicles seen in the system area when blowing-dust conditions were in progress. Questionnaires were then mailed to identified drivers. Of the 1776 questionnaires mailed out, 40% were returned and provided valuable insights into the dust problem from the driver's point of view.

Analysis of the field data indicated that the display on signs was not always appropriate for the reported condition in the field. The data showed that 60% of the respondents had seen a warning message appropriate for the type of weather encountered. Forty-six percent indicated that the signs helped very much. However, the overall conclusion was that the system needed to be improved.

As a result of the first system evaluation, modifications were made to the sign messages and to the manner in which the system was operated. A re-designed system went into operation on 1 June 1976 and was renamed the Dust Storm Alert System (DSAS). The intent was changed from a desire to provide motorists with specific warnings at specific signs to one in which an area-wide general alert of possible blowing-dust conditions was made by activating all 40 signs. The effectiveness of the redesigned system was evaluated from June to September of 1976 by analysis of questionnaires (1174 of them). Seventy seven percent of respondents believed that the information given in radio messages and by signs would help them to avoid a dust-storm accident. The second evaluation by questionnaires suggested that the redesign was worth the effort.

After these two evaluations were completed, several studies were conducted by personnel at ADOT between 1976 and 1988. In 1982, the 40-sign system was cut back to ten signs, and then to four signs in 1984 (Arizona Republic, 2 and 5 Sep 1988). Today, all signs are de-activated, although the original 40 radio-controlled sign boards are still in place. Costs for dismantling would be $10 000. Therefore, gradual cuts in the system have been made since 1976 as a result of two types of study repeated several times over the last twelve years: (a) questionnaires to drivers; and (b) statistical studies of accident rates due to blowing dust. The results of the first kind of study showed that drivers desire some kind of warning, even if only from static signs near the area and from radio weather predictions. The results of the second kind of study suggested that dust storms have apparently been on the decline in the late 1970s and early 1980s. Why this is so has not been fully determined from these sets of studies, but local officials feel that increased development has eliminated abandoned land areas in critical locations which would generally promote dust during major storm events.

No doubt these studies justify actions to reduce the nature of the warning system for severe dust storms in central Arizona, at least based on data used to support this conclusion. Accident-rate changes over time, and driver needs solicited from questionnaires, support the actions taken. The state (ADOT) is also aware that local land-use conditions along the highway, particularly the disturbed agricultural surfaces, have been shown to be linked to many clusters of accidents in the past (Marcus 1976). In fact, small plots of land immediately adjacent to the highway have been experimentally treated to suppress dust problems.

In retrospect, what had not been studied in the early 1970s was the large-scale connection alluded to earlier between extensive winter vegetative secondary-succession invasion on fields coincident with winter moisture provided to barren desert surfaces (including agricultural abandoned land surfaces) and subsequent decreased probabilities of severe dust generation during the summer period in central Arizona. This connection (Figure 7.14) was analysed by Brazel and Nickling (1987) based on long-term data from 1948 to 1983, wherein a sufficient sample size of years was available to develop such a relationship. Applying this concept to the central Arizona region (both Yuma and Phoenix are shown) results in a possible climate cause for the reduced accident rate in the late 1970s and early 1980s. A precipitation/PDSI time plot helps to support this idea (Figure 7.15). Unfortunately, as the dust-warning system did not remain intact throughout the whole period from 1976 to the present, a full long-term evaluation of its effect on dust-accident rates is not possible.

The largest drought period (coincident with high numbers of dust storms) occurred between 1970 and 1972, just before the dust-warning system went into effect. After that time dust storms were never again as frequent, since moisture in central Arizona was generally on the rise from the late 1970s. Therefore the dust warning system was never 'tested' under the conditions for which it was developed. After system modifications were implemented in 1976, the climate conditions were not as conducive to massive dust-storm generation as the severely arid period of the early 1970s in central Arizona.

During the summer of 1988 massive accidents again occurred in the Eloy/Casa Grande locale of central Arizona. The January-to-April winter moisture was lacking to develop significant perennial vegetative ground cover on barren fields, and vast plots of land were again exposed to the vagaries of the summer monsoon winds. The dust-accident problem returned with a vengeance to central Arizona, as did USA nation-wide drought in 1988. These conditions indicate that no continued downward trend is expected in dust-storm frequencies in central Arizona. Judging on past records, a large variation in dust-

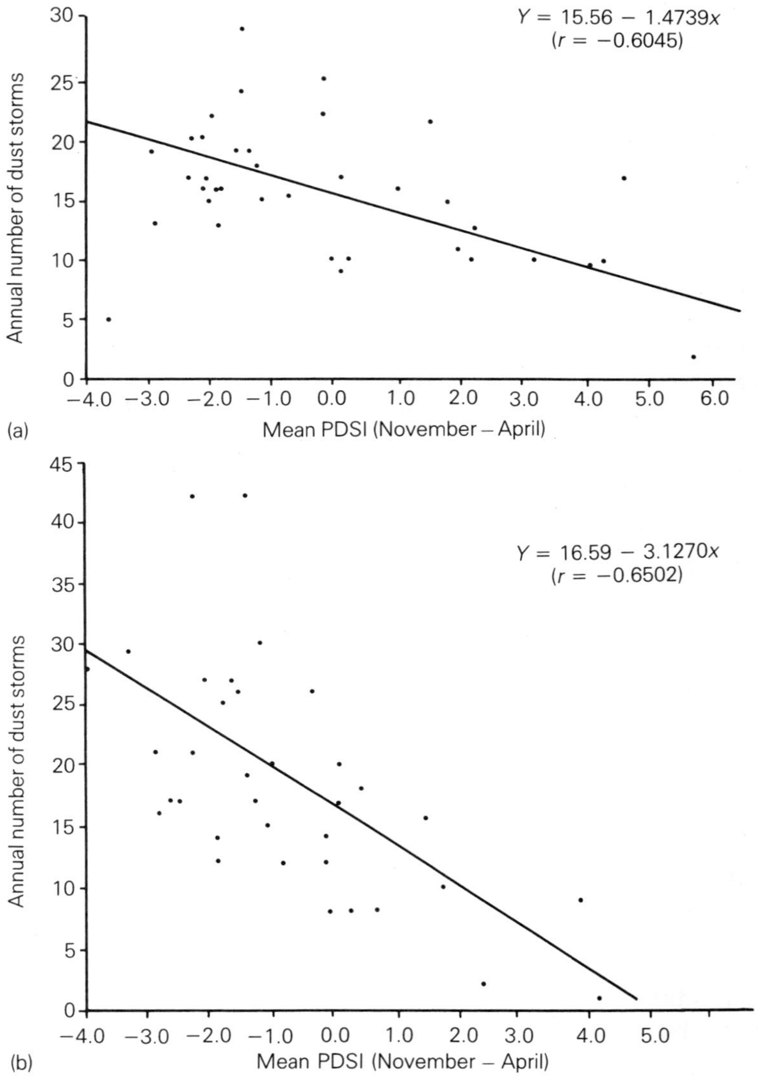

Figure 7.14 Mean Palmer Drought Severity Index (for the period November–April each year) versus frequency of dust events per year for (a) Phoenix, and (b) Yuma weather stations.

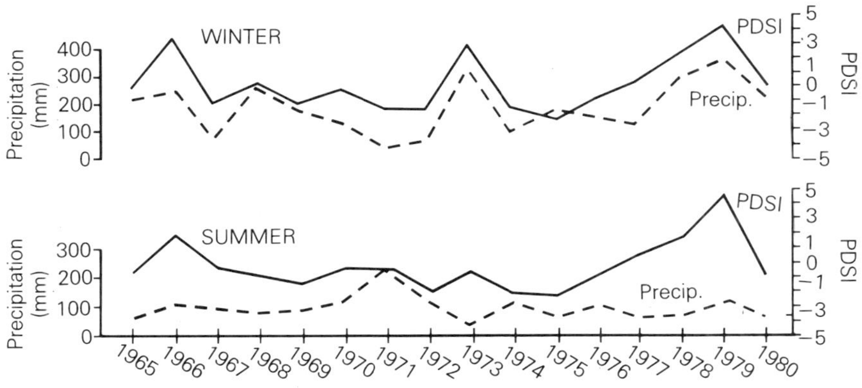

Figure 7.15 (a) Winter (November–April) precipitation totals and mean PDSI, and (b) summer (May–September) precipitation totals and mean PDSI. These values are mean spatial values for a 51 000 km^2 area surrounding Phoenix, AZ: the so-called South Central Climatic Division (modified after Brazel and Nickling, 1987).

storm frequencies is expected over time frames of decades. Eliminating a warning system because of a short-term downward trend in dust problems is evidence of a short-term view of potential future problems.

7.5 FUTURE PROBLEMS OF DUST AND THE DUST HAZARD

The estimation of future effects on the dust hazard in Arizona can only be speculative at best. The possible natural influences may be discussed in terms of a signal of the climate system that is near-periodic or monotonic over time, such as the El Nino Southern Oscillation Index (ENSO) and CO_2-induced climate changes. (The Southern Oscillation is a fluctuation of the inter-tropical general circulation while El Nino is an occasional warning of the western Pacific Ocean which can affect atmosphere conditions world-wide).

El Nino and the Southern Oscillation (ENSO) have been linked dynamically to alterations in climate of the south-west USA in two studies. In the first, Carleton (1987) states that no statistically significant relationship was found between the atmospheric circulation over the south-west USA and the coupled tropical sea surface temperature/ atmospheric index representative of the Southern Oscillation (SOI) for the period 1945–1984. Significant lag correlations were found between Carleton's summer synoptic circulation index and the SOI for 1945–1963 and 1964–1984. However, the south-west USA circulation-SOI teleconnections were deemed unstable over time, and thus not predictable. Brazel (in press) found insignificant correlations between Carleton's summer synoptic index and dust-storm frequencies for central Arizona sites. However, Andrade and Sellers (1988) report a positive relationship between rainfall in Arizona (March–May and September–November) and moderate-to-strong El Nino events over the time period 1899–1983.

More study is needed to establish the nature of periodic circulation changes regionally and globally and their links to surface-climate changes in dust-prone environments of the south-west USA and Arizona. Strong El Nino generation is quasi-periodic, with periodicity of 11 ± 5 years. There may be some predictability between El Nino trends and critical antecedent moisture that affects dust potential in central Arizona.

The clearest signal expected by the year 2065 is the doubled CO_2 and trace-gas warming effect (NAS 1983). Based on prevailing scenarios from model output-gridded temperature and precipitation data of several global climate models, a scenario can be developed on the depletion/supply water ratios for Arizona. State projections on these ratios have been developed not accounting for climate change, but accounting for expected water-consumption changes by the year 2020 (Arizona Water Commission 1977) with high, medium, and low estimates for that year. The most probable ranges of the depletion/supply ratios associated with the ensemble effects of CO_2 warming/drying anticipated for Arizona. In the absence of climate changes, water-use depletion/supply ratios are anticipated to drop due to cutbacks of agricultural land in Arizona have been calculated. With future climate change (+2 °C, −10% precipitation) due to CO_2, the depletion/supply ratios could escalate to 1.23–2.06. Thus further stringent controls on water use may take place.

The implications for dust generation are staggering. Natural climate effects would act to reduce natural vegetative growth on desert surfaces and increase the frequency of drought conditions. Much more land may be retired from agriculture and remain idle. Increased recreational uses of land and construction activity pressures could mean a more severe blowing-dust hazard when wind storms occur. Unknown, of course, is how wind potentials and storm frequencies will vary due to climate change.

The dust hazard must be addressed more intensively in the future. The dust problem is certainly complex and may be partially overcome with careful consideration of land-use planning. Strict air-quality regulations on particulates may in the future represent a negative feedback to the dust hazard in the coming years. However, land near highways and in the surrounding region must be given special attention by federal, state, and local transportation officials and scientists in order to minimize potential adverse effects by future climate. The transportation planners, together with environmental scientists and highway meteorologists, must safeguard the highway systems against the vagaries of future climate changes.

7.6 REFERENCES

Arizona Department of Transportation (1976). *Arizona's Dust Warning System: A Review and Evaluation.* Arizona Dept of Trans. Traffic Operations Services, 22pp. (with appendices).

Andrade, E.R., Jr. and Sellers, W. (1988). El Nino and its effect on precipitation in Arizona and Western New Mexico. *Journal of Climatology*, **8**, 403–410.

Arizona Republic, newspaper accounts of 2 and 5 Sept 1988.

Arizona Water Commission (1977). *Arizona State Water Plan*, Phase II: Alternative Futures.

Brazel, A.J. (in press). Dust and climate in the American Southwest. *NATO ARW* (ed. M. Leinen), Oracle, Arizona.

Brazel, A.J. and Hsu, S.I. (1981). The climatology of hazardous Arizona dust storms. In T.L. Pewe (ed.), *Desert Dust: Origin, Characteristics, and Effect on Man*, Special Paper 186, Geological Society of America, pp. 292–303.

Brazel, A.J. and Nickling, W.G. (1986). The relationship of weather types to dust storm generation in Arizona (1965–1980). *Journal of Climatology*, **6**, 255–275.

Brazel, A.J. and Nickling, W.G. (1987). Dust storms and their relation to moisture in the Sonoran–Mojave Desert region of the South-western United States. *Journal of Environmental Management*, **24**, 279–291.

Burritt, P.E. and Hyers, A. (1981). Evaluation of Arizona's Highway Dust Warning System. In T.L. Pewe (ed.), *Desert Dust: Origin, Characteristics, and Effect on Man*, Special Paper 186, Geological Society of America, pp. 281–292.

Carleton, A.M. (1987). Summer circulation climate of the American Southwest, 1945–1984. *Annals of the Association of American Geographers*, **77**(4), 619–634.

Changery, M.J. (1983). *A Dust Climatology of the Western United States*, National Climatic Data Center, 25pp.

Clements, T. *et al.* (1963). *A Study of Windborne Sand and Dust in Desert Areas.* Tech. Report ES-8, US Army Natick Laboratories, Earth Science Division, 61pp.

Douglas, A.V. (1974). *Cutoff Lows in the Southwestern United States and Their Effects on the Precipitation of This Region.* Final Report, NOAA Contract 1-35241, Tree Ring Lab, University of Arizona, Tucson, AZ.

Fors, J.R. (1977). Tropical cyclone KATHLEEN. *National Weather Digest*, **2**(3), 6–20.

Gillette, D.A. (1981). Production of dust that may be carried great distances. In

T.L. Pewe, (ed.), *Desert Dust: Origin, Characteristics, and Effect on Man*. Special Paper 186, Geological Society of America, pp. 11–27.

Gillette, D.A., Adams, J., Muhs, D. and Kihl, R. (1982). Threshold friction velocities and rupture moduli for crusted desert soils for input of soil particles into air. *Journal of Geophysical Research*, **87**, 9003–9015.

Goudie, A.S. (1978). Dust storms and their geomorphological implications. *Journal of Arid Environments*, **1**, 291–310.

Goudie, A.S. (1983). Dust storms in space and time. *Progress in Physical Geography*, **7**, 502–530.

Hales, J.E. (1974). Southwestern United States summer monsoon source – Gulf of Mexico or Pacific Ocean? *Journal of Applied Meteorology*, **13**, 331–342.

Hales, J.E. (1975). A severe southwest desert thunderstorm: 19 August 1973. *Monthly Weather Review*, **103**, 344–351.

Hales, J.E. (1977). On the relationship of convective cooling to nocturnal thunderstorms at Phoenix. *Monthly Weather Review*, **105**, 1609–1615.

Hecht, M.E. and Reeves, R. (1981). *The Arizona Atlas*, Office of Arid Land Studies University of Arizona, Tucson, AZ. 164pp.

Helgren, D.M. and Prospero, J.M. (1987). Wind velocities associated with dust deflation events in the Western Sahara. *Journal of Climate and Applied Meteorology*, **26**, 1147–1151.

Idso, S.B. *et al.* (1972). An American haboob. *Bulletin of the American Meteorological Society*, **53**(10), 930–935.

Ingram, R.S. (1972). *Summer Duststorms in the Phoenix Area*. Arizona NWS Technical Memorandum: AZ 1.

Karl, T.R. and Knight, R.W. (1985). *Atlas of Palmer Drought Severity Indices (1931–1983) for the Contiguous United States*, National Climatic Data Center, Historical Climatology Series, No. 11, Asheville, NC, USA.

Karpiscak, M.M. (1980). *Secondary Succession of Abandoned Field Vegetation in Southern Arizona*. PhD. Dissertation, University of Arizona.

Lougeay, R. *et al.* (1987). Monitoring changing desert biomass through video digitization of Landsat MSS data: an application to dust storm generation. *Photogrammetric Engineering and Remote Sensing*, **53**(9), 1251–1254.

Marcus, M.G. (ed.) (1976). *Evaluations of Highway Dust Hazards Along Interstate Route 10 in the Casa Grande–Eloy Region*, Final Report, Arizona Dept. of Transportation and Research Paper No. 3, Center for Environmental Studies, A.S.U., Tempe, AZ.

National Academy of Science (1983). *Changing Climate – Report of the Carbon Dioxide Assessment Committee*. US National Research Council, National Academy Press, Wash., DC, 496pp.

Nickling, W.G. and Gillies, J.A. (in press). Emission of fine grained particulates from desert soils. In M. Leinen (ed.), *NATO ARW*, Oracle, AZ.

Nickling, W.G. and Gillies, J.A. (1986). *Evaluation of Aerosol Production Potential of Type Surfaces in Arizona*, EPA Contract No. 68-02-388, MND Associates, 84pp.

Orgill, M.M. and Sehmel, G.A. (1976). Frequency and diurnal variation of dust storms in the contiguous USA. *Atmospheric Environment*, **10**, 813–825.

Palmer, W.C. (1965). *Meteorological Drought*. Research Paper 45, US Weather Bureau, US Dept. of Commerce.

Pewe, T.L. (ed.) (1981). *Desert Dust: Origin, Characteristics, and Effect on Man*. Special Paper 186, Geological Society of America, 303pp.

Wilshire, H.G., Nakata, J.K. and Hallet, B. (1981). Field observations of the December 1977 wind storm, San Joaquin Valley, California. In T.L. Pewe, (ed.), *Desert Dust: Origin, Characteristics, and Effect on Man*, Special Paper 186, Geological Society of America, pp. 223–232.

Chapter Eight

Road Accidents And The Weather

J.P. Palutikof

The total cost of road accidents in Great Britain in 1989 is estimated at about £6 billion (Dept of Transport *et al.* 1990). This figure is calculated from the direct costs of ambulance and police time, hospital treatment and loss of earnings, plus a notional sum for the pain, grief and suffering involved. Clearly the costs to society and to the individuals are enormous, and all reasonable effort should be directed towards reducing the number and severity of road accidents wherever possible.

This chapter examines the relationship between road accidents and weather. It attempts to quantify and cost the proportion of accidents due to bad weather, and looks at the role of weather-information systems in accident prevention. There is no simple relationship between accidents and weather (Andrey and Olley, 1991). In certain circumstances, such as snow conditions, people drive more slowly and carefully and, where possible, postpone or cancel their journeys (Palutikof 1983). This leads to a reduction in the total number of accidents and in the number of serious accidents per unit distance travelled. In wet conditions, conversely, the number of accidents increases (Brodsky and Hakkert 1988). The situation is complicated by a host of other factors which include the longer hours of darkness in winter and the greater volume of traffic on certain days of the week and at certain times of day.

We begin with a general survey of road accidents in Great Britain: the types of vehicles involved, the conditions under which accidents occur and the people most likely to be involved. Statistics from this country are compared with those from overseas.

8.1 STATISTICAL SOURCES

The primary sources of statistics on road accidents in Great Britain are the various publications of the Department of Transport with the

Scottish Development Department and the Welsh Office. These organizations produce a number of annual and quarterly publications which include *Road Accidents Great Britain*, *Transport Statistics Great Britain*, *Road Accidents Scotland* and *Road Accidents Wales*. In addition, the *Monthly Digest of Statistics*, produced by the Central Statistical Office, includes tables of accident statistics. These publications provided the basic information for this study, along with occasional papers produced by the Transport and Road Research Laboratory.

The collation of road-accident statistics is based on the form Stats 19 which is filled out by the police for all road accidents involving human injury or death. This form includes details of the vehicle type, location of the accident, light, weather and road surface conditions and the cause of the accident where this was clearly related to factors such as skidding and/or impact with an object on or off the carriageway. These forms are sent by the 51 police forces of Great Britain to the Department of Transport, who receive some quarter of a million each year.

A road accident is defined by the Department of Transport as follows: 'one involving personal injury occurring on the public highway (including footways) in which a road vehicle is involved and which becomes known to the police within 30 days of its occurrence. The vehicle need not be moving and it need not be in collision with anything . . . Damage-only accidents are not included . . . ' (Dept of Transport *et al*. 1988). On some occasions in this chapter, where it seemed appropriate or where accident figures were not available, data for casualties have been used rather than data for road accidents. A casualty is defined as: 'a person killed or injured in an accident . . . One accident may give rise to several casualties'. (Dept of Transport *et al*. 1988).

For the purposes of any serious scientific enquiry, it is important to determine whether all serious accidents, as defined above, are always reported. To all intents, this is true of fatal accidents. However, a 1979 study in a large hospital casualty department revealed that from a sample of serious injuries sustained in road accidents, 21% had not been reported to the police. This proportion rose to a massive 59% for cyclists involved in road accidents (Dept of Transport 1988). Whereas it is no doubt the case that an indeterminate number are eventually reported within the 30-day limit, it must remain true that road accidents are under-reported. This fact should be borne in mind when considering the results reported in this chapter.

In Great Britain, road-accident statistics were first collected at the national level in 1909. At that time, there were 101 000 registered motor vehicles and 1071 fatal accidents. The 1987 figures were 21 907 000 vehicles and 4694 fatal accidents. This represents a 217-fold increase in vehicles as against only a four-fold increase in fatalities.

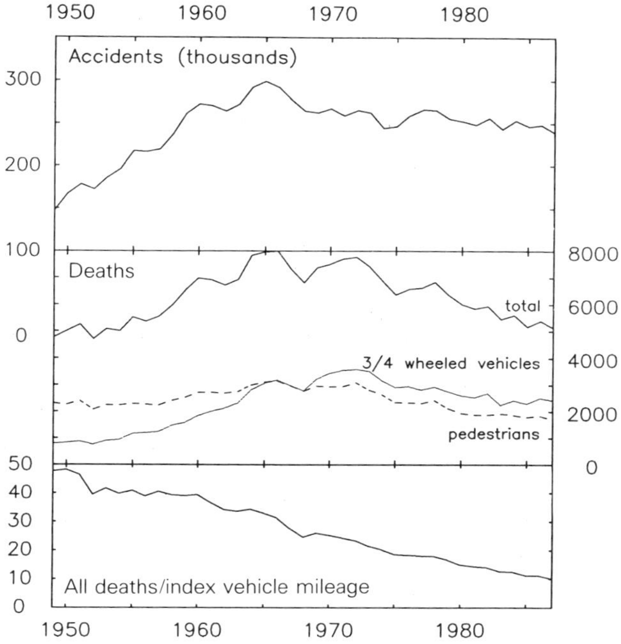

Figure 8.1 Number of road accidents and deaths, Great Britain 1949–1987. (Dept of Transport *et al.* 1988).

8.2 PATTERNS OF ROAD ACCIDENTS

8.2.1 Annual trends

The change over time in the number of accidents on roads in Great Britain has shown two distinct phases over the last four decades. As shown in Figure 8.1, from the start of the record in 1949 the number of accidents gradually increased from around 150 000 per year to a maximum of nearly 300 000 per year in 1965. The next three years showed a sharp decline followed by, since around 1968, a much more gradual decrease. The current figure stands at around a quarter of a million accidents per year. These trends have been interrupted on a number of occasions, for example in 1983 with the introduction of the seat-belt law.

Figure 8.1 also shows the change with time in the number of deaths arising from road accidents for different classes of road user. As already noted, this is a more reliable figure given the under-reporting of serious injuries and, presumably, accidents. The peak year for all road deaths is

1966, which coincides broadly with the worst year for all accidents. However, there is a subsidiary maximum in 1972, related to the fact that the number of deaths for users other than pedestrians, cyclists and motorcyclists did not peak until that year.

It is not really possible to determine if weather conditions have any effect on these annual statistics. Over the period of record there have been a number of individual severe seasons, the influence of which might be expected to show at the annual level. The winter of 1962–63, for example, was for many areas of England and Wales the coldest since records began (Shellard 1968). Heavy snowfalls were recorded. We might expect these conditions to restrict travel, and therefore the number of accidents. Figure 8.1 does show a decline in the number of road accidents and deaths in 1962, followed by a recovery in 1963 to around the 1961 level. Closer inspection reveals that the decline in deaths was due largely to a reduction in the number of pedestrian deaths (and also cyclists and motorcyclists, not shown in Figure 8.1). Deaths in cars, heavy goods vehicles (HGVs) etc. increased in 1962 compared to 1961, although the size of the increase was less than in the preceding years. These results are inconclusive, and could easily be ascribed to some other cause.

When the number of accidents is standardized by dividing each annual total by the index of vehicle mileage for that year, we find that the number of deaths per unit distance travelled has gradually declined with time. It can only be assumed that this is due to improvements in road engineering, car design and public awareness.

8.2.2 Seasonal trends

Not only is there a long-term trend in accident statistics, but the data also display seasonal cycles. Figure 8.2 shows the mean seasonal cycle of road casualties computed for the five-year period 1983–1987. Total casualties are low at the start of the year in the winter months of January and February. They rise gradually towards a peak in the holiday months, then fall slightly in September before rising to the annual maximum in October and November. Of the winter months (December, January, February), December is by far the most severe. Deaths on the road (as a subset of the number of casualties) follow a very similar pattern, with a minimum number in February and a maximum in October–November. Figure 8.2 also shows the total number of casualties per 100 million vehicle kilometres travelled. There is a clear contrast between the first two-thirds of the year, when casualty numbers remain at or below 110 per 100 million vehicle-kilometres, and the last three months when they rise to over 115 per 100 million vehicle-kilometres.

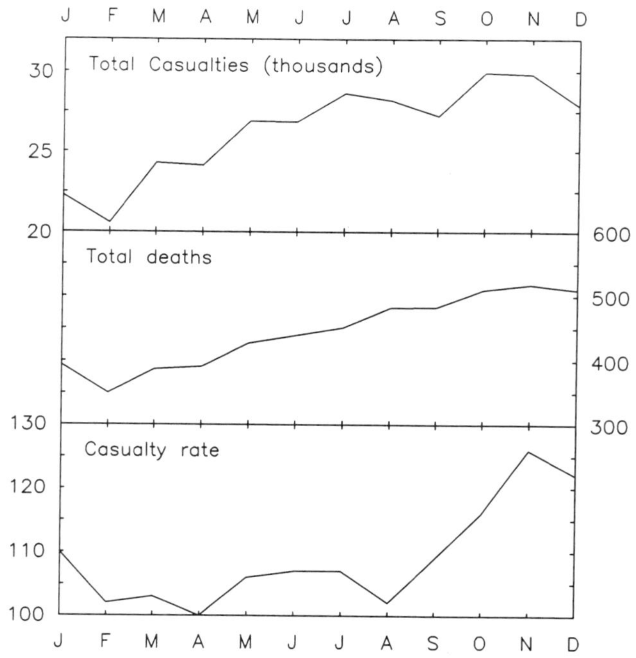

Figure 8.2 Seasonal cycle of number of road casualties, deaths and casualty rate per 100 million vehicle-kilometres, based on 1983–1987 statistics for Great Britain (Dept of Transport *et al.* 1984–1988).

The seasonal cycle varies according to geographical location. As shown in Figure 8.3, areas of the country with an important tourist industry, such as the South-west, show a pronounced concentration of accidents during the holiday months of July and August. For the year shown here, 1985, the July value shares the distinction with November of being the worst month of the year. By contrast, areas such as Greater London, which people prefer to leave in summer if they can, have a very obvious maximum in November and the August figure is in fact below average for the year.

In order to determine the influence of severe seasons, the road casualty statistics for the winter of 1962–63 and the summer of 1976 are plotted in Figure 8.4. The monthly trend of casualties in 1963 is rather different from that for the more recent period. However, it agrees well with the mean statistics calculated from the period 1961–1965, which suggests that the intra-annual distribution of road casualties has changed somewhat since then. There is a clear deficit in road casualties

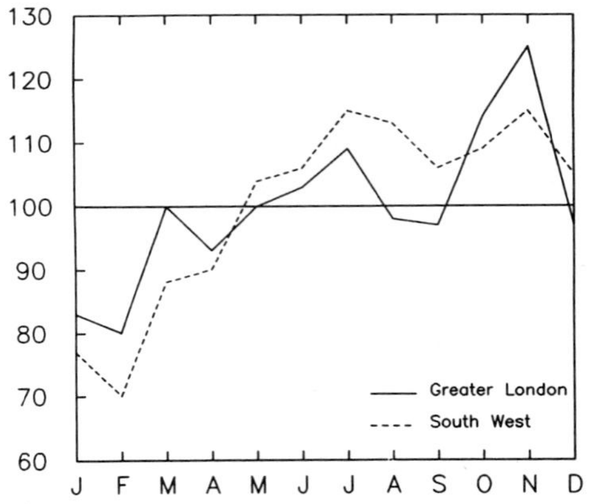

Figure 8.3 Seasonal cycle of all road accidents in Greater London and the South West for 1985. Accidents expressed relative to base (100) equal to average number of accidents per month for the region (Dept of Transport 1987).

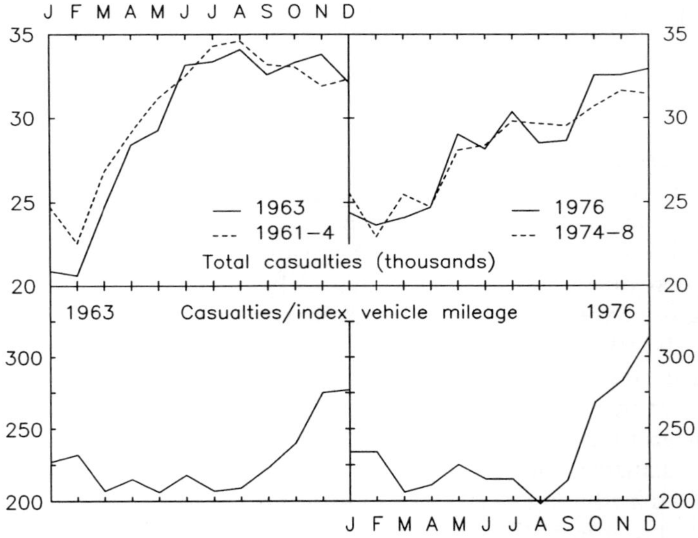

Figure 8.4 Seasonal cycle of all road casualties in Great Britain for 1963 relative to 1961–1965 and for 1976 relative to 1974–1978. (Central Statistics Office, 1962–1966; Central Statistics Office 1975–1979).

in the first two months of the year relative to the mean, which can probably be related to the severe weather. This finding is supported in a paper by the author which showed that a negative relationship exists between the number of casualties in road accidents and an index of snow depth (Palutikof 1983).

Casualties per unit distance travelled are also shown in Figure 8.4. This graph is not exactly comparable to the casualties per 100 million vehicle-kilometres shown in Figure 8.2, since here vehicle distance is expressed as an index relative to a base year, in this case 1958. However, it can be seen that the shape of the curve is broadly similar to that for 1983–1987, suggesting that although the distribution of total casualties through the year may have changed, this is only due to some shift in travel patterns.

The period May 1975 to August 1976 was the driest sixteen months in the UK since records began in 1727 (Doornkamp and Gregory 1980). However, the severity was greatest in 1976 (Smithson 1980) and the fine weather resulted in a significant increase in traffic volumes, according to the Dept of Transport *et al.* (1988). We might expect that the increased traffic might have led in turn to an increase in casualties. Figure 8.4 shows that from May to July total casualty numbers were slightly above the mean for the five-year surrounding period 1974–1978. However, August figures for 1976 were well below the five-year mean, although the drought did not break until September. Any evidence for the effect of exceptionally fine weather on casualties is therefore at best only tentative.

8.2.3 Weekly/daily trends

Casualties sustained in road accidents are more common on certain days of the week and at certain times of the day. In 1986, the most dangerous day of the week was Friday, followed closely by Saturday (see Figure 8.5). In fact, in some years and some areas the number of casualties on Saturday exceeds those on Fridays. The total number of casualties for Monday–Thursday was below 45 000 per day and Sunday, the safest day, had less than 40 000 casualties. Again, however, Sunday casualty figures will sometimes exceed those of Monday–Thursday. As a general rule, we can classify days into two types: safe days (Monday–Thursday and Sunday) and dangerous days (Friday and Saturday).

A study of the five worst days and the five best days for accidents in 1987 (Vickers 1988) found that four of the best days occurred between 13 and 18 January. This period was marked by exceptionally cold weather with heavy snow over much of England and Wales. High winds caused drifting and brought traffic in East Anglia and Kent virtually to a

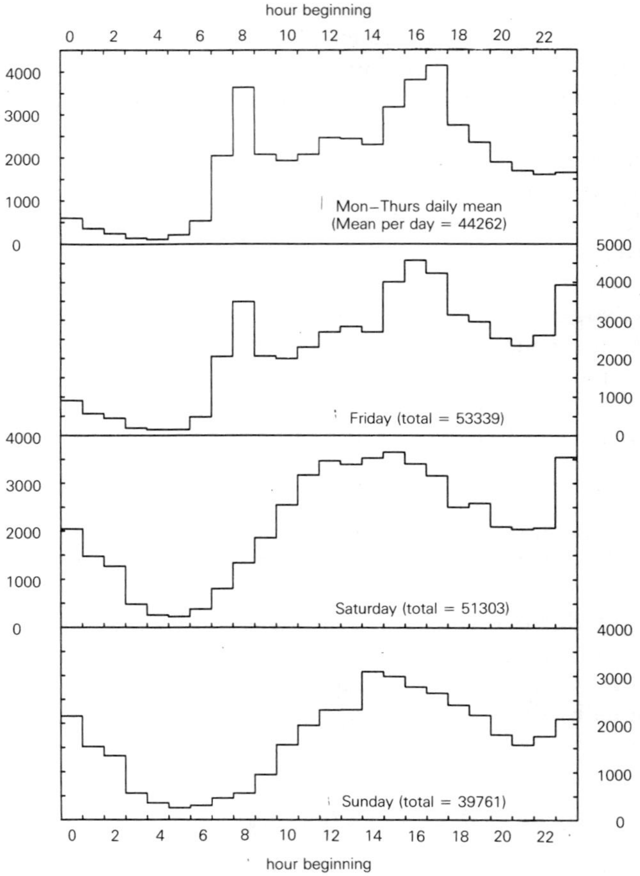

Figure 8.5 Weekly and daily trends in the number of all road casualties, Great Britain 1986 (Dept of Transport *et al.* 1987).

standstill (Anon 1987a). The fifth day was Boxing Day. The worst days were in October (7 and 9 October, well before the 'Great Storm' of October), November (1 day) and December (2 days).

The diurnal distribution of casualties varies from day to day. Weekday casualties show two sharp peaks, at 0800 h and 1600–1700 h (see Figure 8.5). Saturday casualties show a broad maximum in the early hours of the afternoon and again a very sharp upsurge at 2300 h. Sunday is broadly similar but the surge at pub-closing time is much less pronounced.

8.2.4 Comparison with other countries

Road casualty statistics for the UK can be compared favourably with those from other nations, although it should be borne in mind that differences in recording procedures do exist. In a comparison based on 1986 figures, it was found that out of a selection of 26 developed nations, only two countries (Switzerland and Norway) had fewer road deaths per 10 000 motor vehicles, and the UK had the fewest deaths per 100 000 population. However, where pedestrian road deaths are concerned the UK has less reason for self-congratulation. Only Greece, the Irish Republic, Spain, Austria, Czechoslovakia, Hungary, Poland, Yugoslavia, Australia and New Zealand exceeded the 3.3 pedestrian deaths per 100 000 population recorded in this country (data taken from Dept of Transport *et al*. 1987).

8.2.5 Summary

So far we have considered the broad characteristics of road accidents in Britain: the historical perspective, the geographical distribution and the seasonal, weekly and diurnal trends. There are clear hints that weather conditions play a role, although not always as one might expect. Fewer accidents occur in the winter months of January and February, because road conditions frequently deter people from travelling. The total number of accidents, and the number controlled for distance travelled, rises sharply towards the end of the year. Weather conditions are frequently unpleasant in the late autumn and early winter, so that we may surmise that this type of wet, sometimes foggy, weather leads to the end-of-year increase. In the next section we look in detail at weather and road conditions recorded at the time accidents occurred, in an attempt to discover the direct influence of weather on road accidents.

8.3 WEATHER AND ROAD CONDITIONS

Table 8.1 shows the percentage of all accidents occurring under different weather and road conditions in 1987 (taken from data in Dept of Transport *et al*. 1988). Road accident data in this form are only of limited use, because of the lack of any reference point. They tell us nothing about whether wet roads, for example, increase the probability of a road accident, since we have no information about the overall frequency of wet conditions. Table 8.1 does say something about the relative probability of experiencing an accident on different types of road under

Table 8.1 *Percentage of road accidents by weather and road condition, Great Britain 1987*

	Weather condition			
	Fine	*Raining*	*Snowing*	*Fog*
Road type				
All speed limits	81.64	16.40	0.92	1.04
Non-built-up	78.48	17.85	1.44	2.23
Built-up	82.72	15.88	0.73	0.06
Motorways	76.78	19.44	2.02	1.76

	Road condition		
	Dry	*Wet/flood*	*Snow/ice*
Road type			
All speeds	62.84	34.48	2.66
Non-built-up	53.71	41.13	5.13
Built-up	65.62	32.44	1.92
Motorways	60.24	36.54	3.11

Built-up roads are those with houses, shops, etc., close to the carriageway

different weather/road conditions. The greatest risks are associated, understandably, with non-built-up roads (speed limits over 40 mile/h – 65 km/h) and motorways, where speeds are generally higher and there may be no lighting. A higher percentage of accidents occurred on motorways than on non-built-up roads when it was snowing or raining but non-built-up roads were more dangerous when it was foggy or the road surface was wet, icy or covered in snow. The differences are small but persist in the 1986 data, not presented here.

Vickers (1988) in his study of the five best and five worst days for accidents in 1987 found that over 50% of accidents on the best (fewest-accident) days occurred when there was snow, frost or ice on the road, whereas on the worst days this figure fell to 5%. Around 69% of accidents on the worst days occurred on wet roads, compared to 37% on the best days and an annual average of 34%. It should be recalled that two of the worst days fell in October, on 7 and 9. For much of Britain October 1987 was one of the wettest months of the century. Both 7 and 9 October experienced heavy rain in places, and on 9 October more than 50 mm of rain fell in parts of southern England (Anon 1987b). Brodsky and Hakkert (1988) present a valuable review of the existing literature on the risk of road accidents in rainy weather. This paper demonstrates that a number of authors have found a clear statistical relationship between

Table 8.2 *Percentage of road accidents by daylight/darkness, weather and road condition, Great Britain 1987*

	Weather condition							
	Daylight				*Darkness*			
	Fine	*Raining*	*Snowing*	*Fog*	*Fine*	*Raining*	*Snowing*	*Fog*
Road type								
All roads	82.94	15.49	0.80	0.76	78.53	18.58	1.19	1.69
Non-built-up	79.12	17.78	1.35	1.76	77.07	18.02	1.64	3.28
Built-up	84.23	14.69	0.62	0.45	79.02	18.80	1.00	1.18
Motorways	76.42	20.35	1.64	1.59	77.65	17.24	2.94	2.17

	Road condition					
	Daylight			*Darkness*		
	Dry	*Wet/flood*	*Snow/ice*	*Dry*	*Wet/flood*	*Snow/ice*
Road type						
All roads	67.22	30.74	2.02	52.33	43.43	4.21
Non-built-up	57.33	38.38	4.27	45.73	47.20	7.04
Built-up	70.20	28.42	1.36	54.39	42.31	3.29
Motorways	62.83	34.69	2.38	53.97	41.02	4.89

rain and an increase in road accidents, although figures vary on the actual size of the increase.

The data from Table 8.1 are repeated in Table 8.2, but in this case divided according to whether the accident occurred in the daylight or in darkness. On all road types the risk of having an accident in poor road conditions increases substantially in the darkness. In general, poor weather conditions also appear more likely to be associated with accidents in darkness than in daylight. However, the differences are smaller, and in the case of motorways a smaller percentage of accidents occurred in rain after dark than in daylight.

8.3.1 Skidding and road accidents

A large proportion of accidents that occur in unfavourable road conditions can be attributed to skidding. The statistic that is normally used to quantify this factor is the skidding rate, that is, the number of accidents in which one or more vehicles is reported to have skidded expressed as a percentage of all accidents (Hosking 1986). The skidding rate can be computed for different road conditions, which gives an indication of the additional risk when the road is, for example, wet or icy.

Table 8.3 *Skidding rates for different types of vehicle and road surface condition, Great Britain 1987*

	Dry	*Wet or flood*	*Snow and ice*	*All*
Pedal cycles	1.86	4.36	10.71	2.59
TWMV	12.65	27.70	55.34	17.68
Cars	10.40	17.77	46.85	14.15
Buses and coaches	3.59	12.68	33.47	6.57
LGV	9.09	18.14	41.70	13.43
HGV	13.79	20.88	26.30	17.14

Source: Department of Transport *et al.* 1988

Table 8.3 shows the skidding rate for different road-surface conditions and for different classes of vehicle in Great Britain in 1987. The skidding rate for snow and ice is, with the exception of HGVs, around double that for wet and flooded surfaces, which in turn is around double that for dry conditions. (It should be borne in mind, however, that many more accidents take place on dry roads than they do on any other kind of surface: see Table 8.1). Certain classes of vehicle are much more vulnerable to skidding than others. Under all types of road conditions, cyclists have the lowest skidding rates. In snow and ice, two-wheeled motor vehicles (TWMVs) and cars are the most vulnerable, whereas on wet and flooded roads the highest skidding rates are for TWMVs and HGVs. It is of interest to note that HGVs appear to be the least affected by changing road conditions: they have the highest skidding rate for any class of vehicle for dry conditions, but the increase between dry and wet/flooded conditions is only 7% and between dry and snow/ice conditions only 12.5%. Apart from cyclists, HGVs have the lowest skidding rate for snow and ice.

Figure 8.6 shows the geographical distribution, by county, of skidding rates on wet and flooded roads in 1989. This pattern will depend on such factors as geography, traffic speeds and volumes, and road-surfacing materials. Patterns will vary from year to year, but Hosking (1986) looked at five years of regional data for 1979–1982 and found that the South-East and East Anglia had the highest wet-road skidding rate whilst the North, Yorkshire and Humberside and the West Midlands had the lowest rates. These patterns are broadly confirmed by Figure 8.6, but high rates are also found in the South-West, North Wales, North and East Scotland and the East Midlands. The skidding rate for snow and ice is shown in Figure 8.7. Low rates are characteristic of central and southern England, rising towards coastal areas and the North.

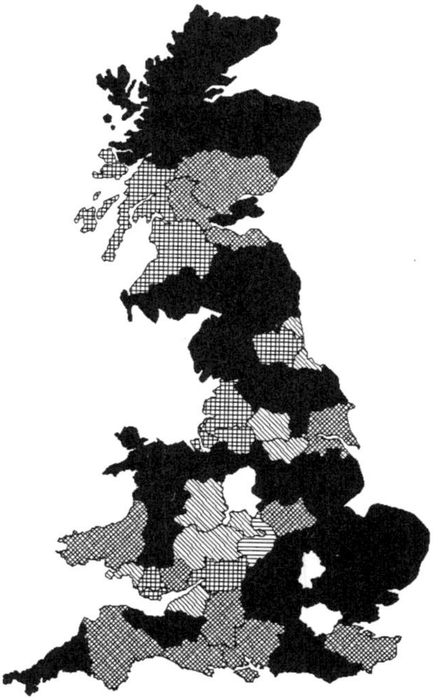

Figure 8.6 Percentage of all accidents on wet or flooded roads involving skidding, Great Britain 1989 (Dept of Transport *et al.* 1990).

These records of the skidding rate can be used to develop a method for calculation of the additional weather-induced cost of road accidents. This method is described in the next section.

8.4 THE COST OF WEATHER-RELATED ROAD ACCIDENTS

As we have already discussed, the fundamental problem with attempting to determine the added weather-induced cost of road accidents lies in establishing a benchmark. Knowing the number of accidents that occurred in poor conditions is not enough, because a certain number of those accidents would have happened irrespective of the weather, for unconnected reasons. One way to approach the problem, and probably the most satisfactory, would be to compare the frequency of accidents in bad weather with the frequency of such weather conditions. Such an approach might work well if the exercise was restricted to one short stretch of road. A manned meteorological station could be set up, and

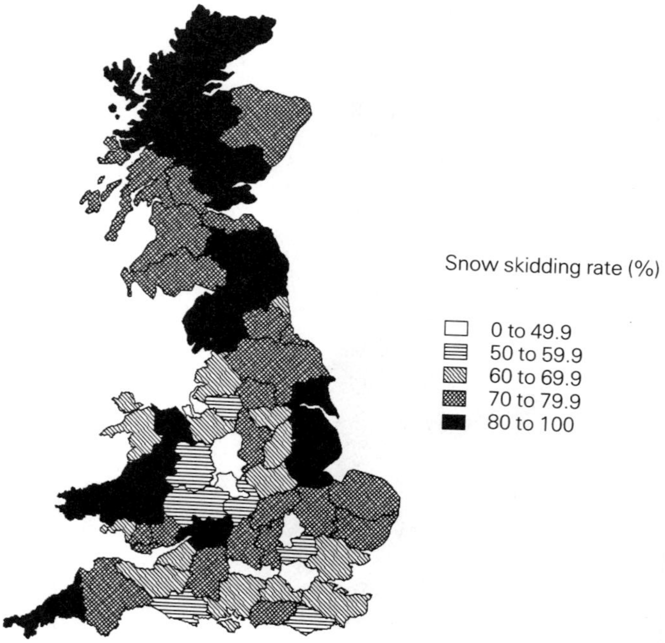

Figure 8.7 Percentage of all accidents on snow-covered or icy roads involving skidding, Great Britain 1989 (Dept of Transport *et al.* 1990).

the required variables, such as visibility and the duration and intensity of rain and snow, measured. However, we are attempting here to assess the added costs for a whole country. The network of meteorological stations is insufficiently dense for this approach, and indeed the costs of setting up the required network would be unjustified.

In preparing the material for this chapter, an attempt was made to link the meteorological data from Heathrow Airport with bad-weather accidents in the Greater London area, with the expectation that the results might tell us something about Great Britain as a whole. A similar exercise was carried out by Smith (1982) who examined weather-related accidents in Glasgow in the light of meteorological data from a nearby site. By looking at daily accident figures he was able to estimate the added costs of accidents due to rain and wet roads at £4.5 million. However, in the study of the Greater London area only annual accident statistics were employed, broken down according to the weather or road condition at the time of the accident. If it could be shown that the proportion of accidents occurring in hazardous conditions was greater than the proportion of time occupied by these hazardous conditions,

Table 8.4 *Observed and expected rate of road accidents in poor conditions, Great Britain 1987*

	Skidding rate (%)	*Total no. accidents*	*Accidents involving skids*	*Expected total no. accidents*	*Difference*
Dry	15.5	150 568	23 338	150 568	0
Wet or flood	29.7	82 481	24 497	68 620	13 861
Snow or ice	63.6	6 363	4 047	2 741	3 622

then we would have an indication of the added risk. The attempt failed, largely because meteorological variables are not measured in a suitable form. In general, statistics are available for a 24-hour period: we can discover, for example, whether fog was observed during that time, whether it snowed, or how much rain occurred. For this exercise, it is necessary to know what proportion of the 24 hours was taken up by such events, and such data are not readily available.

Therefore it was necessary to take another approach. This makes the assumption that the difference between the dry-road skidding rate and the skidding rate on wet, flooded, icy or snow-covered roads is due entirely to the poor condition of the road surface. As such, it cannot take account of accidents not involving skidding, or of other weather-related accidents caused, for example, by poor visibility in fog, snow or driving rain. The example presented here is for 1987. The results will vary from year to year according to the severity of the weather: it has already been noted, for example, that October 1987 was an unusually wet month.

Table 8.4 shows the overall skidding rate for Great Britain in 1987 for the three different categories of road condition (Dept of Transport *et al.* 1988). The dry road skidding rate of 15.5% provides the benchmark. If 15.5% of accidents in dry conditions involved skidding, then if the road condition had no effect we would expect only the same proportion of accidents in wet/flooded or snow/icy conditions to involve skidding. In fact the proportion is much higher. Let us examine the implications of this for wet or flooded roads. From Table 8.4, 24 497 accidents in 1987 involved skidding on wet/flooded roads. Thus the number which did not involve skidding is 57 984. Under dry road conditions, this figure would represent 84.5% of all accidents, since the benchmark skidding rate is 15.5%. Thus the expected total number of accidents under wet/flooded conditions, where the road surface condition had no added effect, is 68 620, given by scaling up from 84.5% to 100%. The difference provides the number of added accidents involving skidding and caused

Table 8.5 *Accidents by road class and condition, Great Britain 1987*

	Motorways	*Non-built-up roads*	*Built-up roads*	*Total*
Dry				
Number	3 329	28 665	118 214	150 208
%	2.22	19.08	78.70	100.00
Wet or flooded				
Number	2 019	21 950	58 439	82 408
%	2.45	26.64	70.91	100.00
Snow or ice				
Number	172	2 738	3 457	6 367
%	2.70	43.00	54.30	100.00

Table 8.6 *Distribution of added accidents by road class, Great Britain 1987*

	Motorways	*Non-built-up roads*	*Built-up roads*	*Total*
Wet or flooded	340	3 692	9 829	13 861
Snow or ice	98	1 557	1 967	3 622

by the road conditions, in this case 13 861 accidents. The same sum is carried out to arrive at the added accidents for snow/icy conditions.

The Department of Transport assigns a cost to a road accident, depending on the class of road on which it occurs, and on the severity (McMahon 1988). These costs contain a material element, for the cost of police, ambulance and hospital services, loss of earnings etc., and a nonmaterial element for pain and suffering. We have to apportion the added accidents according to the road class and the accident severity. Data are available for the number of road accidents by road class and road condition (Dept of Transport *et al.* 1988). These have been used to calculate the percentages given in Table 8.5. This shows that the distribution of accidents between motorways, non-built-up and built-up roads varies little between dry and wet/flooded conditions. However, almost as many accidents occur on non-built-up roads as on built-up roads when the road is snow-covered or icy. It should be noted that the total number of accidents in the final column of this table should agree with the total in column 2 of Table 8.4. In fact there are differences, which may in part be due to the exclusion from Table 8.5 of accidents where the road class is unknown, and may in part be due to rounding errors. The percentage figures in Table 8.5 have been used to prepare Table 8.6,

Table 8.7 *Accident severity by road class and condition, Great Britain 1987*

	Wet or flood		*Snow or ice*	
	Number	%	*Number*	%
Motorways				
Fatal	65	3.22	3	1.74
Serious	399	19.76	36	20.93
Slight	1 555	77.02	133	77.33
All	2 019	100.00	172	100.00
Non-built-up roads				
Fatal	788	3.59	67	2.45
Serious	6 076	27.68	700	25.57
Slight	15 086	68.73	1 971	71.98
All	21 950	100.00	2 738	100.00
Built-up roads				
Fatal	836	1.43	23	0.67
Serious	12 076	20.66	664	19.21
Slight	45 527	77.91	2 770	80.12
All	58 439	100.00	3 457	100.00

where the added accidents from Table 8.4 are distributed by road class.

Table 8.7 gives the breakdown of 1987 road accidents by road condition, road class and severity. Accidents tended to be the least severe on built-up roads, no doubt a function of lower speeds. The proportion of fatal and serious accidents was lower on motorways than on non-built-up roads. For a given road class, the severity was rather less under snow/icy conditions than under wet/flooded conditions.

Table 8.7 forms the basis for Table 8.8, which shows the added accidents by severity. This table is constructed on the assumption that skid-related accidents will have the same severity distribution as all accidents. This assumption was necessary since no published information on the severity of accidents involving skidding could be found. Also shown in Table 8.8 are the costs per accident type, taken from McMahon (1988). This paper estimates that the total cost of accidents in 1987 was £4990 million. Of this amount, 55% was incurred through accidents on built-up roads, 41% on non-built-up roads and 5% on motorways. It should be noted that the categories of accident – fatal, serious and slight – cover a very wide range of circumstances, and therefore the cost of each accident can differ widely from the average

Table 8.8 *Distribution of added accidents by severity, Great Britain 1987*

	Cost per accident (£)	Wet or flood		Snow or ice	
		Number	Total cost (£)	Number	Total cost (£)
Motorways					
Fatal	683 620	11	7 519 820	2	1 367 240
Serious	22 240	67	1 490 080	20	444 800
Slight	2 910	262	762 420	76	221 160
All		340	9 772 320	98	2 033 200
Non built-up roads					
Fatal	575 820	132	76 008 240	38	21 881 160
Serious	22 820	1 022	23 322 040	398	9 082 360
Slight	2 660	2 538	6 751 080	1 121	2 981 860
All		3 692	106 081 360	1 557	33 945 380
Built up road					
Fatal	526 630	141	74 254 830	13	6 846 190
Serious	18 040	2 031	36 639 240	378	6 819 120
Slight	1 560	7 657	11 944 920	1 576	2 458 560
All		9 829	122 838 990	1 967	16 123 870
Total			238 692 670		52 102 450
Total all added accidents = £290 795 120					

figure given in Table 8.8. In particular the category 'serious' can range from severe general shock requiring medical treatment to death more than 30 days after the accident.

McMahon (1988) presents figures on the average cost per accident, allowing for damage-only accidents, in 1987. On built-up roads the cost was only £16 410, compared to £34 140 on non-built-up roads and £37 320 on motorways. Although more accidents occur on built-up roads, as already noted they tend to be less severe and therefore the average cost is lower. Accident costs on motorways are very high, but they tend to occur less frequently: motorways carry 15% of our traffic but account for only 2% of accidents involving injury or death.

Using the cost per accident-type in column 1 of Table 8.8, and the distribution of the added accidents by road type and severity, it is possible to compute the cost of accidents added because of unfavourable road conditions. It can be seen from column 5 of Table 8.8 that the added costs for snow and ice are very much lower than those for wet and flooded conditions. The costs for built-up roads are greater than those for non-built-roads, which in turn are very much greater than those for motorways. The total cost of all added accidents in 1987 is estimated to have been just under £300 million.

It must be borne in mind that a number of assumptions were made in the procedures described above, such as that skid-related accidents carry the same costs as all accidents. Therefore the figure of £300 million can only be taken as an approximation. Moreover, the analysis takes no account of weather-related accidents arising from other causes, for example poor visibility. Strictly, if we were attempting to evaluate the total annual cost of accidents due to bad weather, we should set against this figure the reduction in accidents which occurs because people cancel their journeys in severe winter weather. However, given the limitations of the data, estimation of the cost of all weather-related accidents is not feasible. What we have achieved is to assign costs to an identifiable set of accidents clearly related to inclement weather. In the next section we look at strategies to reduce the number of accidents occurring in bad weather conditions.

8.5 ACCIDENT-PREVENTION STRATEGIES

Some accidents will always happen, due to such factors as driver error, poor visibility, and mechanical failure of the vehicle, acting alone or in combination. However, as Figure 8.1 clearly showed, it is possible to reduce the number of accidents which occur. With respect to those accidents which are in some way linked to poor weather or road surface conditions, there are four possible ways to reduce their number:

(a) road-engineering solutions, including lighting, the geometry of the road and road surface materials;
(b) car-engineering solutions;
(c) road-user awareness and education, not simply in the art of defensive road use, but also in the value of weather forecasts and local radio broadcasts; and
(d) changes to the wider environment.

We examine each of these in turn.

8.5.1 Road-engineering solutions

Sabey (1980a) estimated that one-fifth of all accidents could be prevented through low-cost engineering solutions. These solutions were listed as changes to the geometrical design (especially junction design and control), changes to the road surfacing texture, improvements in road lighting, and modification of road design, traffic management and land use in urban areas. These are described as low-cost measures, which applied nationwide would cost in the region of £100 million. Spread over a number of years this should, according to Sabey, produce a rate of return of £250 million from accident saving.

There are a number of problems with this assessment of the benefits arising from road-improvement schemes, which are clearly illustrated in the analysis of Adams (1988). It is possible to argue that drivers will simply increase their speeds on improved roads, with an unchanged accident level. Injuries may be transferred from one class of road user to another if, for example, lamp-posts are moved to the inside of the pavement to reduce fatalities in cars that swerve off the road. Assessments of the success of 'black spot' treatment (where road improvements are introduced at a known danger spot) need to take into account the accident-migration phenomenon, where accident numbers decline at the black spot itself but increase on surrounding roads (Adams 1988).

These factors cast some doubt on attempts to evaluate the benefits of road improvements. This applies not only to road accidents in general, but also to those caused by poor weather and road conditions.

8.5.2 Car-engineering solutions

The reservations expressed in the previous section must apply equally to any car-engineering solutions: improvements to, say, braking systems may simply cause people to drive more quickly and/or brake later. Sabey (1980a) considered that savings of around 7% were possible through the more widespread introduction of anti-lock brakes and safety tyres. This paper noted that the cost of accidents (at 1980 values) in which the average car is involved over its lifetime was £705. Discounting this figure at 7% over ten years (the average life of a car) indicates that it was worth spending £578 to improve the safety features on a new car, provided all car involvements are then eliminated. It is clear that there must be other, more cost-effective, approaches to the problem of accident prevention.

8.5.3 Driver awareness and education

The discussion in this section is restricted to people in charge of a vehicle, particularly four-wheeled vehicles, since weather-related accidents appear most likely to be caused by this group. There is no doubt that drivers do find certain driving conditions worrying: in a survey of 852 motorists Sheppard (1975) found that the prospect of motorway driving in bad weather conditions caused particular concern. There are four ways in which drivers can be helped to improve their awareness of risks and to proceed with greater safety in poor conditions:

(a) by eduction and training;
(b) by strong enforcement of existing legislation;
(c) through the introduction of new legislation; and
(d) by the provision of appropriate information prior to and during the journey.

(a) Education and training

In a study which examined the statistical relationship between a range of human factors (including age, socio-economic group etc.) and driving performance it was shown that additional driver training is significantly correlated with safe driving habits (Quimby and Watts 1981). Clearly then there is a rationale for further training, but what form should this take? Should it be in the form of a more thorough initial driving test, retraining at periodic intervals, or simply advertising campaigns to alert drivers to specific problems?

Armsby *et al.* (1989) were disappointed to find, in a study of drivers' perceptions of specific hazards, including fog and wet roads, that the drivers were remarkably unaware. Where they did make specific reference to hazards in their interviews, it was generally with regard to the behaviour of other road users rather than to the road environment itself. It became clear that drivers generally 'view the road not as a place at which accident risk varies systematically from place to place, but as a neutral setting in which road users suddenly and indiscriminately threaten their safety'.

This suggests that, in addition to teaching drivers specific skills, such as how to manage a skid, there is a considerable need to educate their perception of the road environment. Drivers should be made aware of the need to make judgements regarding not only the behaviour of other road users but also the condition of the road environment: is the road lit, is the road surface wet/icy, what are the potential risks arising from these conditions? The need is for defensive driving not only with regard to other road users, but also with regard to the environment itself. The education of drivers in this regard can be carried out at all levels: in the initial test, in retraining programmes and through advertising campaigns. It could be done on television as a distance-learning programme. What is required is the will to re-educate drivers in this fashion.

(b) Strong enforcement of legislation

Two pieces of recent legislation are widely regarded as having substantially reduced road accidents: the 80 mg alcohol per 100 ml blood limit imposed on drivers in 1967 and the compulsory front-seat belt legislation of 1983. Estimates of the reduction in accidents due to fear of the breathalyser vary widely, but an authoritative source places the initial reduction in casualties at 11% (Sabey 1980b). This piece of legislation is interesting because we may expect that it has led to an overall reduction in the total number of accidents. This is not necessarily the case with the seat-belt law. This piece of legislation is thought to have saved 7000 fatal or serious casualties amongst front-seat drivers and passengers (Scott and Willis 1985). However, fears have been expressed that, because it increases the driver's sense of security, it has led to more injuries by cars to other road users, particularly cyclists and pedestrians.

Strong enforcement of legislation is not simply a matter of having more police out on the roads looking for people to prosecute. Indeed, this may in some senses be counter-productive: it has been suggested by Adams (1988) that the increase in the injuries: fatalities ratio since 1930 may at least in part be due to the fact that nowadays fewer non-fatal accidents go unrecorded because we have more than double the number of policemen in Britain today compared to 1930. Enforcement may be more a function of social attitudes. Although the breathalyser legislation came into force in 1967, it is only fairly recently that drunken driving has come to be widely viewed as socially unacceptable.

Enforcement may also be quite a subtle thing. At an anecdotal level, we have all travelled through motorway contraflow systems with imposed speed limits of 50 mile/h or lower. And we are all aware that where the speed restriction is accompanied by police radar warning signs, the traffic speed drops to the required level, whereas when there are no such signs the traffic continues to move at speeds well above the limit. There may be no radar trap for miles: simply the warning is sufficient.

There is little doubt that enforcement of legislation saves lives. This is true of accidents that occur because of poor weather conditions, just as it is of any other kind of road accident. A drunken driver exercises less skill, and is more likely to be affected by poor visibility caused by driving rain, for example, than if he or she were fully alert. A driver who feels that the probability of being caught drunk-driving has reached an unacceptably high level, and that this will lead to disapproval in the workplace, home and social circle, is likely to remain sober. The same level of policing and social control needs to be exercised over other driving behaviour linked to accidents, such as excessive speeding.

(c) Introduction of new legislation

The most obvious example of possible new legislation designed to reduce the number and severity of road accidents would be a tightening of the drunk-driving law. The 80 mg of alcohol per 100 ml of blood has been in force since 1967, and it appears that public opinion would be tolerant of a lower limit. Other areas of legislation which we read and hear about frequently and which should be effective in accident limitation, in fine and bad weather, include use of rear seat-belts (which 1991 legislation in the UK has required, where the belts are already fitted) and restricting the access of cars and HGVs to congested inner city areas.

(d) Improving driver awareness

The average vehicle driver on a long journey has little or no access to information about conditions ahead. We read and hear much in the media about developments such as 'intelligent' cars, which will exchange information with vehicles coming from the opposite direction and present it on a dashboard screen. Such innovations are to be applauded, since they should help to prevent, for example, multiple accidents in fog. However, their widespread adoption must lie some distance in the future. In the meantime, it should be possible to improve driver awareness substantially, within the constraints of modern technology.

Before commencing a long journey in the winter season, a driver is likely to telephone some form of forecasting service to obtain information on expected weather/road conditions ahead. Such information is provided by organizations such as the UK Meteorological Office with their regional Weathercall forecasts (Meteorological Office 1989). If the forecast is a bad one, it is quite likely that the journey will be postponed or cancelled. These actions show a level of awareness of the weather and its associated risks that is generally absent throughout the remainder of the year. However, forecastable episodes of heavy rain and fog, which carry their own risks for traffic, can occur in any season. If drivers can be persuaded to use the forecasting services for road traffic at all times of the year, they will then drive with a better knowledge and, more importantly, awareness, of the risks that lie ahead. An aware driver should be a safer driver.

Once the journey has started, it is still possible to pass information to drivers through the radio and roadside signs. Local radio is a useful source for warnings on hold-ups and weather/road conditions. The problem is to persuade people, particularly those from outside the region who may not know the right frequency, to re-tune their radios. Roadside signs announcing the name and frequency of the local station may be useful in this regard. However, in a survey it was found that although 17% of drivers attempted to re-tune their radios after seeing such signs, fewer than half were successful (Owens 1985). Clearly drivers feel a need for information on local conditions which could be met, at least in part, by the provision of roadside signs and car radios which are easy to re-tune. More information on national wavelengths would be useful.

On motorways electronic signs are installed which can be activated to inform drivers about speed limits, whether advisory or restrictive, which the police feel are required because of, for example, poor visibility or an accident. Cross and Parker (1980) found in a survey of driver opinion that only 45% always complied with speed restrictions of this type. The majority of drivers questioned thought it would be helpful to give the reason for the restriction. Certainly it would be possible to do this, giving the nature of, and distance to, the hazard. This would be a worthwhile innovation if it persuaded drivers to proceed in a more responsible manner. In a 1977 study of motorway signals along the M1 in Bedfordshire, Lines (1981) found by observation of the traffic flow that the best response was for restrictive limits imposed because of poor visibility (this combination accounted for 26% of sign settings).

The Transport and Road Research Laboratory has experimented with roadside signs which flash a warning to motorists who are either exceeding the speed limit or following the car in front too closely. In general the response to these was found to be rather poor (Helliar Symons and Wheeler 1984; Helliar Symons and Ray 1986).

It is clear that a number of largely unexplored opportunities exist to provide drivers before and during their journey with information on their own driving standards and on the road environment in which they drive. However, in addition to providing these facilities, considerable effort would have to be directed towards persuading drivers to use this information in a constructive way.

8.5.4 Changes to the broader environment

This category covers those factors which are not *directly* related to traffic on the roads. Perhaps the best example of such an external influence is the possibility of introducing British Summer Time throughout the winter months in the UK. This would have the effect of transferring an hour of daylight from the early to the late part of the day. Table 8.2 showed clearly that the risk of a weather-related accident increases substantially in the darkness. Hillman (1989) suggests that the introduction of all-year-round BST would help to depress the evening accident peak which we noted in Figure 8.5, and states that the reduction at this time would far outweigh the inevitable increase in early-morning accidents. Savings in the cost of road accidents of around £50 million arising from the introduction of all-year-round BST were estimated.

8.6 CONCLUSIONS

This chapter has examined in some detail the effect of bad weather on road accidents. It has been estimated that the cost of road accidents in

1987 was in the region of £5 billion (Dept of Transport *et al.* 1988). Of this amount, we have calculated in this chapter that accidents involving skids caused by bad weather/road conditions alone accounted for some £300 million.

The relationship between bad weather and the number and severity of road accidents is not a simple linear one. Whereas severe winter weather tends to depress the number of road accidents because people tend to postpone or cancel their journey, the number of accidents in rain and on wet/flooded roads increases. Other variables affect this pattern: for example, the risks from rain and wet roads are increased in darkness.

A range of options exist to planners and road engineers attempting to reduce the number of weather-related accidents. Of these, probably the least-explored are those designed to enhance and harness drivers' awareness of the risks attached to poor driving conditions. Much could be done to encourage drivers to make more informed use of forecasting services such as the Met. Office Weathercall. Drivers would pay greater heed to warning signs imposing speed limits along motorways if these gave some indication of the reason for the warning. Wider use could be made of such signs.

The weather makes substantial additions to the road-accident statistics of Great Britain. However, there are statistical difficulties involved in separating out weather-related accidents from the total and, furthermore, we recognize that the weather is a phenomenon beyond our control. For these reasons, weather has perhaps been somewhat overlooked as a cause of road accidents whose influence we can take steps to reduce. There is little doubt that, within the technological and financial constraints which exist, strategies could and should be implemented to reduce the toll of weather-related accidents.

8.7 REFERENCES

Adams, J.G.U. (1988). Evaluating the effectiveness of road safety measures. *Traffic Engineering and Control* **29**, 344–352.

Andrey, J. and Olley, R. (1991). Relationships between weather and road safety: past and future research directions. *Climatological Bulletin*, Ottawa (in press).

Anon (1987a). Weather log January 1987. Looseleaf insert in *Weather*, **42**(3).

Anon (1987b). Weather log October 1987. Looseleaf insert in *Weather*, **42**(12).

Armsby, P. *et al.* (1989). Methods for assessing drivers' perception of specific hazards on the road. *Accident Analysis and Prevention*, **21**, 45–60.

Brodsky, H. and Hakkert, A.S. (1988). Risk of a road accident in rainy weather. *Accident Analysis and Prevention*, **20**, 161–176.

Central Statistics Office (1962–1966). *Monthly Digest of Statistics*, **196–244**. London: HMSO.

Central Statistics Office (1975–1979). *Monthly Digest of Statistics*, **355–408**.

London: HMSO.
Cross, I.G. and Parker, D.B. (1980). Motorway matrix signalling. *Journal of the Institution of Highway Engineers*, **27**, 5–9.
Department of Transport (1987). Road accident statistics English regions 1985. *Statistics Bulletin (87)*, **34**. London: Department of Transport.
Department of Transport *et al.* (1987). *Road Accidents Great Britain 1986: the Casualty Report*. London: HMSO.
Department of Transport *et al.* (1988). *Road Accidents Great Britain 1987: the Casualty Report*. London: HMSO.
Department of Transport *et al.*, (1984–1990). *Road Accidents Great Britain 1983–1989*. (After 1984 Subtitled *The Casualty Report*). London: HMSO.
Doornkamp, J.C. and Gregory, K.C. (1980). Assessing the impact. In Doornkamp, J.C., Gregory, K.J. and Burns, A.S. (ed.), *Atlas of Drought in Britain 1975–76* pp 79–80 London. Institute of British Geographers.
Helliar Symons, R.D. and Wheeler, A.H. (1984). Automatic speed warning signs – Hampshire trials. *TRRL Laboratory Report*, **63**, Crowthorne: Transport and Road Research Laboratory.
Helliar Symons, R.D. and Ray, S.D. (1986). Automatic close-following warning signs – further trials. *TRRL Laboratory Report*, **63**, Crowthorne: Transport and Road Research Laboratory.
Hillman, M. (1989). More daylight, less accidents. *Traffic Engineering and Control*, **30**, 191–192.
Hosking, J.R. (1986). Relationship between skidding resistance and accident frequency: estimates based on seasonal variation. *TRRL Research Report*, **76**, Crowthorne: Transport and Road Research Laboratory.
Lines, C.J. (1981). The effect of motorway signals on traffic behaviour. *TRRL Supplementary Report 707*, Crowthorne: Transport and Road Research Laboratory.
McMahon, K. (1988). Cost of accidents. In *Road Accidents Great Britain 1987: the Casualty Report* produced by Dept of Transport, Scottish Development Dept and Welsh Office, pp. 35–36. London: HMSO.
Meteorological Office (1989). *Annual Report 1988* (Met. 0.988) London: HMSO.
Owens, D. (1985). Driver response to experimental roadsigns which display local radio station frequencies. *TRRL Research Report 47*. Crowthorne: Transport and Road Research Laboratory.
Palutikof, J.P. (1983). The effect of climate on road transport. *Climate Monitor*, **12**, 46–53.
Quimby, A.R. and Watts, G.R. (1981). Human factors and driving performance. *TRRL Laboratory Report 1004*. Crowthorne: Transport and Road Research Laboratory.
Sabey, B.E. (1980a). Road safety and value for money. *TRRL Supplementary Report 581*. Crowthorne: Transport and Road Research Laboratory.
Sabey, B.E. (1980b). The drinking road user in Great Britain. *TRRL Supplementary Report 616*. Crowthorne: Transport and Road Research Laboratory.
Scott, P.P. and Willis, P.A. (1985). Road casualties in Great Britain during the first year with seat belt legislation. *TRRL Research Report 9*. Crowthorne: Transport and Road Research Laboratory.

Shellard, H.C. (1968). The winter of 1962–3 in the UK – a climatological survey. *Meteorological Magazine*, **97**, 129–141.

Sheppard, D. (1975). The driving situations that worry motorists. *TRRL Supplementary Report 129*. Crowthorne: Transport and Road Research Laboratory.

Smith, K. (1982). How seasonal and weather conditions influence road accidents in Glasgow. *Scottish Geographical Magazine*, **98**, 103–114.

Smithson, P.A. (1980). Rainfall. In Doornkamp, J.C., Gregory, K.J. and Burns, A.S. (ed.) *Atlas of Drought in Britain 1975–76*, pp. 15–16. London: Institute of British Geographers.

Vickers, J. (1988). Best and worst day for accidents. In *Road Accidents Great Britain 1987: the Casualty Report* produced by Dept of Transport, Scottish Development Dept. and Welsh Office, pp. 55–58. London: HMSO.

Chapter Nine

International collaboration

Erkki Nysten

International collaboration in meteorology has always been important, because the weather does not obey any regional boundaries. In the fields of road traffic, road maintenance and road construction international co-operation has a long tradition, but international collaboration in highway meteorology only started about two decades ago as part of the COST programme: European Co-operation in the Field of Scientific and Technical Research.

The first co-operation venture was the project COST 30 (electronic traffic aids on major roads), and especially one of its sub-projects called COST 30 Theme 8, studying the relationship between road traffic and weather. During the final stage of COST 30 considerable interest arose among the European countries for further co-operation in the field of road weather. This led to the formation of the Standing European Road Weather Commission (SERWEC), which is a voluntary organization comprising meteorologists and highway engineers from about fifteen countries.

The final report of the COST 30 project recommended further research and development work on road-weather problems. The executive committee of SERWEC undertook the preliminary work of drafting a proposal for a new COST project which was approved in October 1986. COST 309 project on road meteorology and maintenance conditions was officially set up in February 1987.

International co-operation on road meteorology is likely to continue. SERWEC is working actively, and new European scientific and technical research projects will be set up in which road and weather conditions are managed: for example the DRIVE project, which stands for Dedicated Road Infrastructure for Vehicle Safety in Europe. DRIVE is a three year European Community programme aiming to improve road safety and efficiency, reduce vehicle emissions and noise pollution through the application of advanced information technology in road transport.

9.1 THE COST 30 PROJECT

The COST 30 project (electronic traffic aids on major roads), which formed part of the COST programme of European co-operation in the field of scientific and technical research, was set up in 1970. Participating countries were Austria, Belgium, the Federal Republic of Germany, Finland, France, Italy, the Netherlands, Sweden, Switzerland, the United Kingdom, Yugoslavia and the European Communities. The overall objective of the project was to improve traffic safety and flow conditions on major high-speed roads, through the promotion of electronic traffic aids for detection of road conditions and communication with the drivers. The project was specifically aimed at the development of functional specifications that would ensure compatibility across national borders.

The work of COST 30 was divided into nine research themes as follows:

(a) in-vehicle speech communication with the driver;
(b) in-vehicle visual communication;
(c) variable signs and signals;
(d) radio broadcasting of traffic information;
(e) information needs of drivers and road authorities;
(f) automatic or manual detection of incidents affecting traffic;
(g) clear, correct and unambiguous terms for use in messages in different languages;
(h) automatic detection of bad weather conditions;
(i) equipment for control centres and control strategies, data transmission, and proposals for an international demonstration.

9.1.1 COST 30 Theme 8

The aim of Theme 8 was to develop a prototype system to predict, detect and give warning of hazardous changes in visibility, wind speed and gustiness, skid risk caused by weather, and surface water. Further aims were to examine the problem of short-term weather forecasting and its possible applications to the operation of road networks, and to consider ways of identifying weather 'black spots'. Seven signatory countries participated actively in this work; Austria, Finland, Germany, the Netherlands, Sweden, Switzerland and United Kingdom.

The work of Theme 8 was largely concerned with the development of a road-weather detection system comprising a number of localized road-weather monitoring stations. Each station would enable road-weather conditions to be automatically detected on a specific section of road. The system could monitor particular black spots or be extended to cover selected points along the road or even the entire road network. More detailed information than that provided by the existing meteorological services would then become available and drivers, road-maintenance crews and traffic-control authorities might all expect to benefit. Using the information from the weather service and the data automatically observed on the road weather stations it would be possible to predict the surface temperature of the road some hours ahead.

The final report of the COST 30 project was published in 1981, with the recommendation for a public demonstration or experiment where the results of the project could be presented in practice. The final report also suggested that work should continue on some of the themes, including Theme 8.

9.2 COST 30 BIS PHASE

The COST 30 BIS project was set up in April 1982. The recommendations contained in the final report of the COST 30 Committee included:

(a) a public demonstration of a motorway control system on an existing motorway site in the Netherlands section south of the Hague; and
(b) the contination of research in the following fields:
  (i) road/vehicle communications: use of microelectronics;
  (ii) use of Citizen's Band radio in European countries;
  (iii) automatic incident detection (AID);
  (iv) regional radio broadcasting of traffic information;
  (v) weather detection;
  (vi) variable traffic signals.

A new management committee was set up to guide the COST 30 BIS programme. It included representatives of all the signatory states, plus a delegation from Spain.

9.2.1 COST 30 BIS Demonstration project

The Management Committee of COST 30 agreed that a demonstration should be staged for the following reasons.

(a) It would provide confirmation of the practical feasibility of applying results already obtained in individual countries in an international setting, with a variety of foreign drivers.
(b) It would ensure the feasibility of integrating components into a complete system.
(c) It would demonstrate the complete system internationally, allowing administrations to review the likely total impact of the system.
(d) It would investigate the effectiveness and provide better estimates of the costs and benefits of an overall system and parts thereof.
(e) It would allow an assessment of public reactions to the system.

The site chosen for the demonstration was the motorway network between Delft and Rotterdam in the Netherlands, where a national control system was already being installed by the Dutch administration. Additional equipment was added to achieve the international demonstration, which was organized by the planning group. The demonstration started in May 1983 and remained in operation for one year.

The main aims of the road-weather service are to improve traffic safety, to reduce traffic costs and maintenance costs and to optimize road maintenance. In order to achieve these aims the road authorities must be provided with specific local and area weather forecasts so that they can be prepared to prevent slippery driving conditions from developing by taking suitable action in adequate time. Also, drivers must be provided with information on weather and driving conditions on the roads via radio, television, telephone and changeable signs. Drivers would get forecasts also for planning travelling times and itineraries, especially in heavy traffic.

In the demonstration the traffic-control system and the weather-detection network were operated separately, partly because of their different communication requirements and partly because the weather information was not used to switch on the signs automatically. The two systems were operated by the same people, however.

Weather detection was achieved by two different systems: the Swiss ice early-warning system and a more complicated system including videotex, which was a combination of Dutch, German, Finnish and Swiss contributions.

The videotex-based weather-service system included two automatic road-weather stations. Both stations were connected to the central computer in the traffic control centre. Three terminals were available for this weather-service system; one for the operator in the Delft traffic control centre, one for the meteorologist at the Rotterdam airport, and one for visitors to the demonstration. The two visibility detectors were connected to the weather stations. The central computer was operated by Finnish software, including warning procedures and a road-surface temperature-prediction model.

The Swiss ice early-warning system included two outstations with sensors, a receiving unit and a display unit. The information from this system was used only by the traffic-control operator.

The weather-detection equipment was used continuously to assist the operators in selecting the appropriate pictograms. For both systems the information was updated regularly and was presented to the operator on a video-terminal. In the case of dangerous weather conditions an audible alarm was given.

In the videotex-based system the weather information from the stations was updated every minute in both graphical and in tabular form. The warning procedures for poor visibility, strong cross-winds and asphalt freezing were automatically run in the system as well. Predicting the surface temperature some hours ahead, a method based on the energy-balance model was used by the meteorologists at Rotterdam airport. The meteorologists were also preparing area forecasts using road-weather data and conventional weather information, to assist the road maintenance people in their operations. Prediction of weather conditions was of particular value for short-term planning of road maintenance.

In the ice early-warning system three alarm levels were used. Alarm level 1 indicated that there was a danger of black ice arising purely from weather conditions. Alarm level 2 indicated that the thawing agent still present on the road would no longer be effective. A danger of black-ice formation would exist if the road temperature were to keep falling. Alarm level 3 meant that ice had actually formed at the detectors, or no more thawing agent remained because of precipitation.

9.2.2 COST 30 BIS Theme 2: Detection of weather conditions

In addition to the demonstration, road-weather problems were also treated under Theme 2 in COST 30 BIS. Participating countries were Austria, Belgium, Federal Republic of Germany, Finland (coordinating) France, the Netherlands, Spain, Sweden and Switzerland. In addition information was obtained from Italy, the United Kingdom, Denmark and Norway.

The aim was to examine the problem of short-term weather forecasts and their repercussions on road maintenance, traffic behaviour, the identification of bad weather 'black spots' and communication with drivers to warn them of dangerous conditions.

Because the work of Theme 2 was based mainly on correspondence the result of this sub-project was more or less limited to ascertaining the state of art in participating countries.

In June 1985, an international seminar on 'Electronics and Traffic on Major Roads' was held in Paris at OECD Headquarters. At this event the final report of the COST 30 BIS project was presented and a discussion about the results and future activities took place among over 100 participants from 20 countries.

9.3 SERWEC: STANDING EUROPEAN ROAD WEATHER COMMISSION

In the closing phase (in February 1984) of the demonstration in the Netherlands, an International Road Weather Conference was held in Delft and in the Hague. The conference was arranged by the committee responsible for COST 30 BIS Theme 2-Detection of weather conditions. This conference was attended by over 50 people from 13 countries. In the discussions about concluding COST 30 BIS, the meeting appointed a voluntary steering committee of seven people, on an *ad hoc* basis, to draw up an agenda for future co-operation. The results from the demonstration, the research work and the conference have proved that there is an urgent need for further research and international co-operation in the field of weather conditions on the roads.

The steering committee met at Birmingham University in September 1984. A constitution for the Standing European Road Weather Commission was agreed. The constitution was approved by the Second International Road Weather Conference, which was held in Copenhagen from 26 February to 1 March 1985. About 90 people from 14 countries attended the conference in Copenhagen. The Third International Road Weather Conference was held in Tampere in Finland in February 1986, with 80 participants from 15 countries, the Fourth Conference took place in November 1988 in Florence in Italy, and the Fifth in Tromso in February 1990.

The following Constitution of the Standing European Road Weather Commission was approved in 1985:

(a) The Commission shall be known as 'The Standing European Road Weather Commission' (SERWEC) and all organizations and individuals with an interest in road weather are eligible for membership.

(b) The Commission shall have as remit the under-mentioned functions and such other functions as the Commission shall from time to time agree:

  (i) The Commission shall operate as a forum for the exchange of information relevant to the field of highway meteorology. This shall include management, maintenance, road safety, meteorology, environmental protection and any other area of interest considered relevant by the Commission.

(ii) From the information collected it shall seek to identify those areas where increased and/or new research and development may yield improvements in practices, techniques, systems and methodology, to the general benefits of the art.

(iii) It shall encourage the undertaking of trials, studies and research into such identified areas.

(iv) A newsletter shall be set up as the official organ of communication, between participating members, to external national and international organizations and to technical publications.

(v) Contact shall be initiated and maintained between the Commission and bodies such as COST, OECD, PIARC, WMO and other bodies so as to ensure that the work of the Commission is recognized.

(vi) The Commission shall ensure, insofar as is possible, attendance at all seminars, conferences and symposia which may prove beneficial to the Commission's purposes.

(vii) An international conference shall be held each year which shall be the Annual General Meeting of the Commission. At this conference an executive committee shall be elected to carry out the tasks defined by the Commission.

(viii) The Commission shall from time to time consider for publication memoranda and other material relevant to its area of interest.

(ix) The Commission shall charge a small membership fee so as to maintain its independence.

At the international seminar on Electronics and Traffic on Major Roads in Paris in June 1985, where the COST 30 BIS project was closed, some statements about plans for future activities were made. At the closing session of the COST seminar, the chairman of the COST Technical Committee of Transport, made the following statement:

(a) There is need for research and development work in the field of road weather conditions, in order to improve traffic safety, and to help winter road maintenance.

(b) Taking into account the resources needed and the need for harmonization and standardization, co-operative research work is highly appreciated.

(c) The following are examples of important research topics:
  (i) detection and short-term forecasting of road surface conditions;
  (ii) dissemination of information on road and weather conditions.

(d) It would be very much appreciated if SERWEC could prepare a proposal for this new research project.

Serwec's steering committee decided at the meeting held in June 1985 to make a proposal (officially by the Finnish delegation) for a new project to COST's Technical Committee of Transport (TCT).

9.4 COST 309: ROAD WEATHER CONDITIONS

The proposal for a new COST project was approved by TCT in October 1986. The Memorandum of Understanding for the implementation of a European Research Project on Road Weather Conditions (COST Project 309) was signed by 11 countries: Austria, Denmark, Finland, France, the Netherlands, Norway, Portugal, Spain, Sweden, Switzerland and the United Kingdom.

When deciding on the COST 309 project the COST Transport Committee set the following objectives for the Project:

(a) to explore the most effective methods of detecting, forecasting and mapping hazardous weather-related road conditions and of taking action to improve their effects;
(b) to quantify the costs and benefits of such methods and actions;
(c) to recommend areas where common research can be carried out and common operational standards established.

The main research subjects of the COST 309 project were defined as:

(a) road weather detection;
(b) road weather forecasting;
(c) road weather statistics; and
(d) road weather service strategies.

The duration of the project was to be three years beginning in spring 1987. The project was to be carried out in three phases. The aim of Phase 1 was to list and evaluate past and current studies related to each of the topics and to define particular projects where co-operative work was required. The duration of Phase 1 was 6 months. During Phase 2, from autumn 1987 to autumn 1989, the management committee of the project would carry out common research projects and have discussions to achieve common standards for the operational exchange of information, and to encourage workshops on subjects of particular interest. Phase 3, the last 6 months, was to be reserved for overall analysis of the results, for production of the final report, and for recommendations on techniques to be used and standards to be adopted.

The management committee of COST 309 decided to carry out research work on several topics coordinated by different countries, as follows:

(a) Sensors and measuring systems (Sweden)

The objective was to develop methods of comparing various sensors and to develop new sensors for measuring road conditions e.g. freezing point, de-icing concentration. Sweden had been testing sensors since 1983 at a test station in Gothenburg and had thus developed expertise in this field.

(b) Detection and prediction of fog (France)

The objective was to develop new sensors for measuring visibility and new methods of forecasting fog.

(c) Overall systems, thermal mapping and data transmission (UK)

The objective was to expand investigations on the overall systems to produce recommendations on the optimum network of sensors suitably sited, to use all available weather and road information to identify road sections where climatic conditions are a major cause of accidents. This could help in planning the installation of warning devices, road-weather stations and the use of de-icing pavements. The objective was also to evaluate the different types of data transmission, their reliability and cost efficiency, and the compatibility of different systems.

(d) Weather radar and satellite information (Spain and Cost 73)

The objective was to investigate the usefulness of weather radar and satellite information for forecasting road-weather conditions and its use by roadmasters. Co-operation is recommended with the COST 73 project which is dealing with the weather radar network.

(e) Prediction of road conditions (Sweden)

The objective was to evaluate methods for short-term prediction (about 12 hours) and for real-time (1–3 hours) of relevant road weather parameters at selected spots on the road network.

The objective was also to expand this evaluation to the forecasting of road conditions and to develop models describing the road-surface conditions as a function of traffic volume, meteorological parameters, climatic conditions and road works. The purpose of such models is to make theoretical analyses of various maintenance inputs (including salting) in order to optimize the resources used.

(f) Weather atlas and winter index (Denmark)

The objective was to develop a standardized weather atlas on a national basis giving information about climatic conditions during winter. This could be a useful tool when comparing methods and resources used for road maintenance in winter in different countries.

(g) Weather and accidents (Norway)

The objective was to establish a standardized practice for recording road accidents, weather conditions and road conditions. This could serve as a useful database when calculating the potential costs and benefits of various measures for road-weather maintenance activities.

(h) Cost-benefit analyses of road weather services (Finland)

The objective was to develop methods and to carry out studies to quantify the costs and benefits of road-weather services to the authorities and road users, including accident costs and environmental costs.

(i) Communication between meteorologists and road-maintenance authorities (the Netherlands)

The objective was to promote better use by road-maintenance personnel, of the information provided by the meteorological service and local weather stations. This topic aimed to include an evaluation of the use of new information systems and the development of training programmes and manuals.

(j) Dissemination of information to road users and standardization of formats (United Kingdom, France and Switzerland)

The objective was to assess the types of information which could be of use to road users, and the technical methods for dissemination, taking the legal aspects into account. The aim was also to enable road users to be aware of road conditions in neighbouring countries. To attain this, there is a need for standardization of road-weather information (both actual and forecast) disseminated between national centres and road users by various media. The topic also includes standardizing the terminology used to describe particular road-surface conditions.

(k) Effects of de-icing agents and other maintenance techniques on road conditions (Management Committee)

The objective was to evaluate the best means of using meteorological data to optimize salting and to determine the effect of de-icing agents under different weather conditions (snow, ice, fog, rain, wind and temperatures).

9.5 OTHER INTERNATIONAL RESEARCH PROGRAMMES

The Organization for Economic Co-operation and Development (OECD) has had a Road Transport Research Programme for a number of years. Reports have been produced on fog and accidents. A new programme on salt usage in winter road maintenance was begun in 1987 concentrating on analysing how salt works and what alternatives there are to its continued use, given the environmental impacts that can result from high salt usage.

In Europe collaborative programmes are moving into a new phase. The DRIVE programme aims to produce by 1995 operational, real-time multi-lingual road information and navigation systems in Europe, and operational integrated traffic management systems in major European cities and corridors (DRIVE Workplan 91, Draft p. 7). Research related to weather hazards does not form a large part of the programme though road condition and weather monitoring systems are included in the many projects for which proposals have been invited. Improvement of communication systems for safety improvements are also the concern of the projects being studied under PROMETHEUS (PROgraMme for a European Traffic system with Highest Efficiency and Unprecedented Safety).

9.6 REFERENCES

COST 30 BIS (1985). Final Report on the Project 'Electronic Traffic Aids on Major Roads'. EUR 9835.

COST 30 BIS. Final Report on the Demonstration Project Group, XII/852/84 CEC

COST 30 BIS. Final Report of Theme 2: Detection of Weather Conditions XII/512/84 CEC.

COST/228/87 CEC.

Memorandum of Understanding for the Implementation a European Project on Road Weather Conditions (COST Project 309).

DRIVE: R&D in Advanced Road Transport Telematics in Europe. Workplan 91. Draft. Brussels, Commission of the European Communities.

Chapter Ten

Conclusions

A.H. Perry and L.J. Symons

In this slim volume consideration has been given to a number of road weather hazards and their alleviation. The main emphasis has been on problems of winter weather in Europe and North America with some divergence to other matters, such as sand blowing, which are of virtually no significance in northern Europe but constitute major problems in some arid lands. It has been shown that design as well as maintenance must have regard to weather and climate. Loss of tyre adhesion and visibility and the blocking of highways are the main ways in which the weather contributes to accidents and much research has been, and is being, devoted to overcoming these hazards but, with constant increase in traffic and in the speed of all classes of motorized vehicles there can be no easing of effort or complacency over the need for ever more refined design and improved maintenance procedures, which must be preceded by adequate research, monitored in operation and modified as shown to be necessary. Safety on the roads cannot, unfortunately, be assured by any international or other programme, all that can be done is to provide the best practicable environment in man-made terms to make good use of the natural environment and reduce its adverse impact.

Highway meteorological work is essentially of an interdisciplinary nature and one of the most encouraging developments of the past decade has been the immensely improved collaboration between scientists and engineers, with significant contributions from geographers, psychologists and other specialists who were hardly involved with highway and other transport hazards two decades ago. Meteorologists and climatologists, especially, have become much more involved in both theoretical and practical terms. The World Meteorological Office has provided a venue at which road problems could be discussed along with other applications of meteorology in society (Perry and Symons, in *World Meteorological Office*, 1991). Cost, interdepartmental official accounting and entrenched attitudes, however, continue to handicap the dissemination of information which could add to road safety. Broadcasting of road information by the media is also ill-organized in the UK. Though some improvements have been carried out in recent years there is no systematic national or regional broadcasting of detailed information on road conditions readily obtainable by all road users comparable with that provided as a matter of course in many European countries. This reflects the low esteem in which public service broadcasting is held in the UK, a situation which appears likely to become worse rather than better with government proposals for reorganization of broadcasting on an increasingly commercial basis from 1991.

As highway meteorologists world-wide look ahead into the 21st century, advance planning will have to take account of the changed environmental conditions that global warming is likely to bring. While sensitivity to weather and climatic change is high not all the expected changes would be detrimental. Amongst the beneficial effects might be reduced expenditure on winter maintenance activities in many parts of western Europe, including the UK. In some areas milder winters may extend the pay-back period of investment in ice detection technology. Savings should also be possible in capital expenditure programmes concerned with snow clearing equipment, while decreases in the number of freeze-thaw cycles and of the severity of frost should reduce structural damage to roads and bridges in many places (Perry, 1991). On the debit side more frequent and severe episodes of low level pollution occurring during hot sunny weather in summer can be expected. Climate change would have implications for weather-related road accidents, for example, in the UK a weak connection has been noted over the last few years between summer temperature and accidents reported to the police, probably as a result of less parental control being exerted on children playing outdoors. Projected increases in precipitation in some mid-latitude countries, although small, could exacerbate problems of road flooding, landslips and corrosion of steelwork on bridges. Greater incidence of hot weather would mean more frequent adverse weather for resurfacing and surface dressing. If storminess increases, disruption from fallen trees and from the overturning of vehicles, such as was widespread in NW Europe during severe gales in January – February 1990 could become an increasing problem (Perry and Symons, 1990).

Comprehensive studies of the likely impacts of climatic change on road transport have yet to be carried out and both broad scale and specific research is urgently required.

10.1 REFERENCES

Perry, A.H. (1991). Transport. In *The Potential Effects of Climate Change in the United Kingdom*. Department of the Environment, HMSO, 91–5.

Perry, A.H. and Symons, L.J. (1990). Highway meteorology – an overview of hazards. *Economic and Social Benefits of Meteorology and Hydrological Services*. Geneva: WMO, 436–40.

Index